DETERMINISM AND FREEWILL

ARCHIVES INTERNATIONALES D'HISTOIRE DES IDEES

INTERNATIONAL ARCHIVES OF THE HISTORY OF IDEAS

Series Minor

18

J. O'HIGGINS S. J.

DETERMINISM AND FREEWILL:
ANTHONY COLLINS' *A PHILOSOPHICAL*
INQUIRY CONCERNING HUMAN LIBERTY

DETERMINISM AND FREEWILL

ANTHONY COLLINS'

A Philosophical Inquiry concerning Human Liberty

edited and annotated

with

a discussion of the opinions of Hobbes, Locke, Pierre Bayle,
William King and Leibniz

by

J. O'HIGGINS S. J.

MARTINUS NIJHOFF / THE HAGUE / 1976

ISBN 90 247 1776 0

PRINTED IN THE NETHERLANDS

TABLE OF CONTENTS

PREFACE

The *Philosophical Inquiry concerning Human Liberty* of Anthony Collins' was considered by Joseph Priestley and Voltaire to be the best book written on freewill up to their own time. Priestley admitted that it converted him to determinism and it had a powerful effect on Voltaire in the same direction. It seems important to place in its wider historical context a book which so influenced such men and which greatly impressed the *philosophes* in general. Therefore – and because such an account has value in itself – the Introduction contains a survey of the freewill controversy from the time of Hobbes to that of Leibniz, giving in some detail the opinions of Hobbes, Locke, Pierre Bayle, William King, Archbishop of Dublin, and Leibniz and an account of the Scholastic doctrine of liberty of indifference – opinions which either influenced Collins or against which he reacted.

The value and originality of Collins' works need assessing. He was also at times liable to misinterpret or misunderstand the authorities he quoted. I have, therefore, subjected the *Inquiry* to a detailed critique. This also gives cross-references to parallel passages in Collins' works and those of the authors who influenced him, and, by discussing the philosophical and theological questions to which his writings give rise, obviates the need for a good many footnotes in the notes that follow the text.

INTRODUCTION

DEISM

Anthony Collins (1676–1729) was one of the leading members of that group of men who, in the first half of the 18th century were designated free-thinkers or Deists. They were a very varied collection, ranging from the impecunious and rather disreputable Irishman, John Toland, with his pantheistic leanings, and the ex-tallow chandler's apprentice, Thomas Chubb, to the aggressive if cautious fellow of All Souls College, Matthew Tindal, and the wealthy Essex squire, Collins himself. Their beliefs varied. Samuel Clarke, in his Boyle lectures of 1704–1705 divided them into four classes: those who held the existence of God but denied that he concerned himself in the government of the world; those who held the existence and material providence of God but denied that he was concerned with the morally good and evil actions of men; those who believed in God and his providence and his insistence on man's obedience but denied the immortality of the human soul; and those who believed in God, in providence, in Natural Religion and a future life with rewards and punishments, but who denied Revelation. The denial of Revelation was uniform in all of them. The belief that unaided reason could work out for itself the whole of the contents of religion was their common characteristic. They all held that the existence of God and of his essential attributes can be rationally demonstrated – with some reservations in the case of Toland with regard to the nature of God himself. But they never founded a single school of thought and with regard to one important question, that with which the text here reproduced is concerned, they differed widely. Tindal and Collins were determinists, Thomas Chubb a defender of the freedom of the will. They were regarded by their contemporaries as enemies of the Christian faith, yet Tindal and Chubb called themselves Christian Deists and Collins remained a communicating member of the Church of England

till his death and was angered, apparently genuinely, by accusations that he was a hypocrite in so doing. They were an interesting phenomenon; partly a product of the reaction against the religious enthusiams of the seventeenth century and the religious wars and yet not quite men of the Enlightenment. But they contributed to the Enlightenment. They corresponded in England to the circle that surrounded Comte Henri de Boulainvilliers in France, though their views were less extreme than those expressed in the clandestine manuscripts that were circulating in the latter country.[1] Their writings were reported on and often summarised in the many contemporary literary journals that were published in France and Holland and their works were translated into French – the *Philosophical Inquiry* had two separate French translations. They were the centre of heated theological controversies that raged in England from the time of the publication of Toland's *Christianity not Mysterious* in 1696 to the middle of the eighteenth century. Apparently they got the worse of the exchanges. But they and the attitude of mind which they represented had a considerable influence in England on what G. N. Stromberg describes as a religion that "rested on a good deal of indifference to all except a minimum of moral rectitude"[2] and in France their effect is more easy to trace in Voltaire, D'Holbach and the Encyclopedists. Of all the English writers of his time, Collins was the one who was most noted in the contemporary French literary journals, and in Jacques André Naigeon's volumes on Philosophy in the *Encyclopédie Méthodique* it was he who was given the second longest article, and in that article a translation of his work on free will was reproduced in full.

ANTHONY COLLINS

In social position Anthony Collins was by far the most favoured of the Deists. He was born in Heston, in Middlesex, of a family that came from the Isle of Wight, that had legal connections and that became landowners in Essex. His grandfather, Anthony, had been a Bencher and Treasurer of the Middle Temple; his father, Henry, was called to the bar in 1667, but did not practise. He himself went first to Eton, then, for a year, to Cambridge and finally to the Middle Temple. He was never called to the bar, though his interrupted legal training may have been of use to him in his later polemical writings. He himself considered that his rather

---

[1] For an account of these, cf. I. O. Wade, *The Clandestine Organisation and Diffusion of Philosophic Ideas in France, from 1700 to 1750*, (Princeton, 1938).
[2] G. N. Stromberg, *Religious Liberalism in Eighteenth Century England*, (Oxford, 1954) ,p. 170.

unsystematic education might have been a disadvantage to him. He was a very wealthy man. It was estimated that his father was worth £ 1,800 a year and he himself married the daughter of one of the richest men in London, the banker Sir Francis Child.[3] At first he lived in London – after the early death of his first wife in 1703 – in Lincoln's Inn and then in Lincoln's Inn Fields; then at the fine old mansion of Hunterscomb in Buckinghamshire, where he received Queen Anne and her court who "took delight in walking in his fine gardens" and where he refused the blandishments of the local gentry to stand as parliamentary candidate for the county, presumably in the whig interest; and finally, after short-lived stays in Banstead, Sutton and Whaddon in Surrey, he moved to Essex in 1715. Here, though he still kept a town house, he became very much the country gentleman. At the same time that he was earning notoriety as a freethinking writer, in his own county he was a pillar of county society, Justice of the Peace, county Treasurer and Deputy Lieutenant. His reputation as a writer did nothing to damage his standing as a country gentleman and from a manuscript life in the British Museum, written presumably by a county acquaintance, one would never gather that he had put pen to paper except to sign official documents – and in the official documents contained in the Essex county record office, his is, for this period, the signature that most frequently occurs. He remained Justice and Treasurer till his death in 1729.

WRITINGS

Without doubt the most important and most formative experience in Collins' life was his friendship with John Locke. He may first have met the latter in the sixteen-nineties but the close friendship existed only during the last two years of the philosopher's life. It was exceedingly close. "Why do you make yourself so necessary to me," wrote Locke. "I thought myself pretty loose of the world, but I feel you begin to fasten me to it again."[4] Locke had a high opinion of Collins' ability and may well have been responsible for his turning to a career as an author. "I know no body that understands it" – the *Essay on Human Understanding* – "so well,"[5] he wrote to Collins, and when he mused on the "openings to truth,

[3] Brit. Mus. Addit. MSS. 4282, f. 243. For a fuller account of the life of Collins and an analysis of his library, cf. J. O'Higgins, *Anthony Collins, the Man and his Works*, (*International Archives of the History of Ideas*, The Hague, 1970).

[4] J. Locke, *Works*, (6th ed., London, 1759), III, p. 730. Locke to Collins, 24 June, 1703.

[5] *Ibid.*, III, p. 742. Locke to Collins, 3 Apr., 1704.

and direct paths leading to it" which he thought he saw in his old age, but considered himself too old to follow up, he wrote "it is for one of your age, I think I ought to say for you yourself, to set about it."[6]

Locke died in 1704. The first of Collins' works were published in 1707. In them he showed the strong influence of his friend but he was to move away, finally far away, from Locke's position. In the *Essay on the Use of Reason* of 1707 he took up much the same position as that of Toland in his *Christianity not Mysterious* of 1696. He held that there is no room for mystery in religion. Propositions are either according to reason and fully comprehensible or contrary to reason and there is no room for any other category. Locke admitted truths above reason, though he was rather ambiguous when he came to discuss them, but Collins certainly took Locke's theory of ideas as a starting point for his argument. In the same year, 1707, began a long controversy with Samuel Clarke, on the immateriality and natural immortality of the soul. Taking up Locke's casual remark that we do not know whether "omnipotency has not given to some systems of matter fitly disposed, a power to perceive and think,"[7] Collins, unlike Locke, plainly held that the soul is material and that matter is capable of the power of thought. God, alone, he conceded, is a spiritual being. Clarke got by far the better of the debate, but Collins' pamphlets – four in all – were particularly beloved by the atheist Jacques André Naigeon, who, since he did not read Clarke himself, but only Collins' pamphlets, in translation, maintained that Clarke had the worse of the argument. In 1710 came the *Vindication of the Divine Attributes* denying man's analogical knowledge of God, and, three years later, the most notorious of Collins' writings, *A Discourse of Freethinking*, a defence of freethinking as such, very anti-clerical, that was immediately translated into French and caused a great furore on the continent as well as in England, but that was savagely and effectively dealt with by the great classical scholar Richard Bentley, who demolished the scholarship of Collins' work, and by Jonathan Swift, who satirised its logical weakness. Three years later there followed the *Philosophical Inquiry concerning Human Liberty*, one of the least notorious but probably the most important of Collins' works, and then there was a gap of seven years during which he was most engrossed in local administration and became the Treasurer of the County of Essex.

In 1724 began the last period of his literary activity. In that year his *Historical and Critical Essay on the Thirty Nine Articles* – a book that

---

[6] *Ibid.*, III, p. 734. Locke to Collins, 29 Oct., 1703.
[7] J. Locke, *An Essay Concerning Human Understanding*, IV, 3, n. 6, (Fraser's edition, Dover Publications, New York, 1959), II, p. 193.

was the sequel to a brief pamphlet of 1709 – argued that the first clause of Article 20 of the Thirty Nine Articles – asserting the right of the Church to decide disputes in doctrine and matters of liturgy – was a forgery, inserted, no doubt by episcopal malice and ambition, in the seventeenth century. He certainly succeeded in showing that something remarkably odd had happened to the Article between 1562 and 1571, though he did not prove this charge of forgery. And then came his final theological works, the *Discourse of the Grounds and Reasons of the Christian Religion* (1724), the *Letter to the Author of the Discourse of the Grounds and Reasons* (printed in 1726 and first published in 1737), *The Scheme of Literal Prophecy Considered* (1726) and *A Letter to Dr. Rogers* (1727). This was the most important controversy in which he was engaged. *Grounds and Reasons* received over thirty replies. The works were a direct attack on one of the most popular arguments for the truth of Christianity – that from prophecy. Collins argued that the Messianic prophecies could only have been fulfilled in Christ in a metaphorical or – in his own loose use of the terms – 'allegorical' way and it was plain that he was implying that this meant that there had been no real fulfilment at all. It was ironical that Locke's friend and disciple should so strike at the last and only fundamental and necessary article of belief that was insisted on by Locke, the belief that Jesus is the Messiah. Yet, in spite of this final polemic, Collins died – and the story seems to have good grounds for our belief in its authenticity – saying that he had always endeavoured to the best of his abilities to serve God, his king and his country and that "he was persuaded he was going to that place which God had designed for them that love him" and asserting that "the Catholic religion is to love God and to love man." He was still not a man of the Enlightenment.

THE CONTROVERSY ON FREEWILL

The *Philosophical Inquiry concerning Human Liberty* forms part of a controversy that had excited England since the middle of the seventeenth century. John Bramhall, Bishop of Derry and later Archbishop of Armagh, had met Thomas Hobbes in Paris in the forties and argued with him on the question of freewill. The Marquis of Newcastle persuaded the debaters to put their arguments on paper and in 1654 Hobbes' contribution "*Of Liberty and Necessity*" was published without his consent. Bramhall replied and this in 1656 led to the publication of Hobbes' "*Questions concerning Liberty, Necessity and Chance.*" The controversy continued till 1668, five years after Bramhall's death, ending with Hobbes' reply to Bramhall's *Castigation of Hobbes' Animadversions* and his

Catching of Leviathan the Great Whale, written in 1668 but not published till 1682. Hobbes' *"Of Liberty and Necessity"* had considerable influence on Anthony Collins, and earlier his determinism had aroused the opposition of the Cambridge Platonists, particularly of Ralph Cudworth, in his unpublished treatise *"Of Freewill,"* and of Henry More in his *Enchiridium Ethicum* of 1667. John Locke in his *Essay* dealt with the question of the will in his chapter on "Power", but the British controversy assumed international importance with the publication in 1702 of William King, Archbishop of Dublin's, *De Origine Mali*. Pierre Bayle, the effect of whose writings on Collins was of first importance, had already been obsessed, in his *Dictionnaire Historique et Critique* (1695–1696), with the problem of evil and its implications with regard to the freedom of the will. Now in his *Réponse aux Questions d'un Provincial*, a sequel to the *Dictionnaire*, he devoted nineteen chapters to King's work and four of these to a criticism of King's theory of freewill. Leibniz also, in 1710, published, with his *Theodicy*, a critique both of King's and of Hobbes' works. Leibniz was quoted by Collins and in many ways their theories of freewill were similar, but Collins' basic opinions had been enunciated in his earlier writings, in, for example, his *Essay on the Use of Reason*. Bayle had a real impact on his *Inquiry*, particularly in his interpretation of the phrase "liberty of indifference." Bayle and Locke were the two major influences on Collins' philosophical ideas and also, in his opinion on freewill, Hobbes' *Treatise on Liberty and Necessity*.

THE *PHILOSOPHICAL INQUIRY*

Already in 1707 Collins had shown himself a determinist. In his *Essay concerning the Use of Reason* he dealt with the problem of the reconciliation of the Divine foreknowledge and man's freewill[8] – the subject of William King's sermon to which the *Vindication of the Divine Attributes* was a reply. He solved the problem to his own satisfaction by saying that all things, including human choices, are determined in their causes and as such can be foreseen by an all-knowing God.

There are causes that ever determine the will, as the appearing good or evil consequences; there are other causes of the appearing good or evil consequences, and causes of those causes and so on; and no one action in this long progression of causes, extend it as far as you please, could possibly not happen. ... My action of taking or forbearing this peach is as certain as is the color and flavor which makes it agreeable or disagreeable, or as is the determination of my will according to its appearing agreeableness or disagreeableness.

[8] *An Essay Concerning the Use of Reason*, (London, 1707), pp. 45–50.

This, he says, is the liberty which ought not to be opposed to necessity "but only to compulsion." His basic idea was already clear. He developed it in the work of 1717. This achieved some literary success. It had a second edition in 1717, a third in 1735 and a fourth, published at Glasgow, in 1749; while Joseph Priestley republished it, with a preface, in 1790. A French translation, made by a Huguenot refugee, de Bons, an acquaintance of Collins' life-long friend and protégé, the exile French journalist Pierre Desmaiseaux, was published at Amsterdam in 1720, in Desmaiseaux's *Receuil de diverses Pièces*. In 1754 appeared a quite new translation ("Eleutheropolis," 1754) made by the blind French philosophe Pierre Lefèvre de Beauvray, who apparently was not aware at first of de Bons' translation. De Beauvray added copious notes of his own, attacking Clarke and Chubb and showing his belief both in determinism and the materiality of the soul. Finally it was reproduced in the *Encyclopédie Méthodique* in 1791. In 1717 there was a brief review of the work in the *Acta Eruditorum* of Leipzig. The *Journal de Trévoux* reported it, in its literary news from London, in July 1719, saying that Collins had confused moral and physical necessity and mistaken liberty of indifference for insensibility. The *Journal des Savants* gave it a brief notice in a longer review of the *Receuil*. The *Journal Litéraire* of the Hague gave a very full abstract of the work in 1718. It was reviewed therefore in France, Germany and the Low Countries and in Germany at the university of Tübingen it was taken in 1722 as the subject of two dissertations for the Master's degree, held before Johann Eberhard Roesler, professor of philosophy.[9] In England there was only one reply to the work, that of Samuel Clarke, to whom Collins had referred as really agreeing with his thesis. Clarke said that Collins confused physical and moral necessity and that he failed to distinguish between the judgement of the intellect and the act of choice. He criticised Collins' interpretation of indifference and his saying that motives determine a man. "When we say, in vulgar speech, that motives or reasons determine a man 'tis nothing but a mere figure or metaphor. 'Tis the man that freely determines himself to act."[10]

Collins did not reply to Clarke, apparently because he thought that the latter was threatening him with the civil power by asserting that his book was a danger to society. Unlike Collins' other important works the *Inquiry* did not cause a major scandal in England. It seems to have had more effect on the continent. It was one of the factors that effected the conver-

---

[9] For a fuller account of the reception and effect of Collins' works on the continent, cf. *Anthony Collins*, pp. 201–223.

[10] S. Clarke, *Remarks upon a Book entituled, A Philosophical Inquiry concerning Human Liberty*, (London, 1717), p. 11.

sion of Voltaire to determinism and in England it was decisive in the case of Joseph Priestley. "It was in consequence of reading and studying this treatise," wrote the latter, "that I was first convinced of the doctrine of necessity."[11] Voltaire was fulsome in his praise. In his *Métaphysique de Newton* (1740) he dealt with the controversy between Collins and Clarke over freewill. He wrote "de tous les philosophes qui ont écrit hardiment contre la liberté, celui qui sans contredit l'a fait avec plus de méthode, de force et de clarté, c'est Collins, magistrat de Londres."[12] He was undergoing a crisis in his belief in human liberty. In 1748 when the *Métaphysique* was incorporated in the *Eléments de la Philosophie de Newton*, the objections brought forward by Collins against freewill were among those which were most influential in shaking his belief in liberty. In 1764 in the article "Liberté" in the *Dictionnaire Philosophique* and in 1766 in the *Philosophe Ignorant* his position had become the same as that of Collins. He wrote

cette question sur la liberté de l'homme m'intéressa vivement; je lus les scolastiques, je fus comme eux dans les ténèbres; je lus Locke, et j'aperçus des traits de lumière; je lus le traité de Collins, qui me parut Locke perfectionné; et je n'ai jamais rien lu depuis qui m'ait donné un nouveau degré de connaissance. Voici ce que ma faible raison a conçu, aidée de ces deux grands hommes, les seuls, a mon avis, qui se soient entendus eux-mêmes en écrivant sur cette matière, et les seuls qui se soient fait entendre aux autres.[13]

In its effects at least the *Philosophical Inquiry* was a book of some importance.

THOMAS HOBBES

Collins' work had a considerable influence at least on two of the more important men of the eighteenth century. It itself shows the influence of Hobbes, Locke, Bayle and possibly Leibniz. It is, therefore, of some interest to see exactly what were the theories of these thinkers with regard to freewill.

Thomas Hobbes was a mechanistic determinist. His theory of freewill and of human nature derived from his general theory of causality. For him, philosophy was concerned with the causes and properties of bodies, material bodies and bodies politic. Motion is the "one universal cause"[14] and animal psychology is explained by what he describes as vital motion

---

[11] *The Theological and Miscellaneous Works of Joseph Priestley*, (Rudd's edition, Hackney, 1817–1831), III, p. 457.
[12] Voltaire, *Oeuvres Complètes*, (Molland edition, Paris, 1879), XXII, p. 413.
[13] *Ibid.*, XXVI, p. 55.
[14] T. Hobbes, *English Works*, (London, 1839), I, p. 69.

"the course of the blood, the pulse, the breathing, the concoction, nutrition, excretion, etc. to which motions there needs no help of imagination" and "animal motion, otherwise called voluntary motion."[15] "Sense, in all cases," he says, "is nothing else but original fancy, caused ... by the pressure, that is, by the motion, of external things upon our eyes, ears and other organs thereunto ordained."[16] "Conceptions and apparitions are nothing really, but motion in some internal substance of the head."[17] The effect of this external pressure is to produce animal motions as well as phantasms[18] and these animal motions proceed to the heart and produce pleasure or pain; pleasure if the vital motion of the blood is helped, pain if it is diminished. Good and evil are relative terms. "The object of any man's appetite or desire, that is it which he for his part calleth good: and the object of his hate and aversion, evil."[19] Hobbes, however, does assume that objects have certain qualities which do us good or evil.[20] Man, he holds, naturally moves towards that which augments and away from that which diminishes the vital actions and his every action is causally determined.

That which necessitateth and determines every action ... is the sum of all things, which being now existent, conduce and concur to the production of that action hereafter, whereof if any one thing now were wanting, the effect could not be produced. This concourse of causes, whereof every one is determined to be such as it is by a like concourse of former causes, may well be called (in respect they were all set and ordered by the eternal cause of all things, God Almighty) the decree of God.[21]

Spontaneous actions, done without deliberation, his opponents would readily grant not to be free, but Hobbes denied freedom to those that are performed after election. Animals, children and madmen, he said, often deliberate.[22] "When a man deliberates whether he shall do a thing or not do it ... he does nothing else but consider whether it be better for himself to do it or not to do it."[23] However, he equiparated deliberation with appetite. He described it as "nothing else but alternate imagination of the good and evil sequels of an action, or, which is the same thing, alternate hope and fear, or alternate appetite to do or quit the action of which he deliberateth" and, he said, "in all deliberations, that is to say, in all

---

[15] *Ibid.*, III, p. 38.
[16] *Ibid.*, III, p. 3.
[17] *Ibid.*, IV, p. 31.
[18] *Ibid.*, I, p. 405.
[19] *Ibid.*, III, p. 41.
[20] *Ibid.*, V, p. 192.
[21] *Ibid.*, IV, p. 246.
[22] *Ibid.*, IV, p. 244.
[23] *Ibid.*, IV, p. 273.

alternate succession of contrary appetites, the last is that which we call the will, and is immediately next before the doing of the action, or next before the doing of it become impossible."[24] Judgement is a product of sense[25] and "the last dictate of the judgement, concerning the good or bad, that may follow on any action, is not properly the whole cause, but the last part of it, and yet may be said to produce the effect necessarily, in such manner as the last feather may be said to break a horse's back, when there were so many laid on before as there wanted but that one to do it."[26] Finally he defined liberty as "the absence of all the impediments to action that are not contained in the nature and intrinsical quality of the agent."[27]

This was a system of iron causal determinism based on Hobbes' theory of the world, of motion and matter, and of human nature. How far Collins accepted or studied Hobbes' general philosophy is not to the point. Certainly he inclined to the view that thought is really matter – in the brain – in motion. But of Hobbes' influence on his theory of freewill there can be no doubt. Hobbes' assimilation of appetite to judgement is to be found in Collins. Hobbes' argument that one is determined by what most appeals to one's appetite is that of Collins and their definition of freewill is the same. There are textual similarities as well. The metaphor of the last feather is found in Collins though in a different context. Both give the commendation of the character of Cato of Utica by Velleius Paterculus as an example to prove the point that it is better to be virtuous by nature than by the action of an alleged freewill, and there is one passage, unacknowledged, – in the *Philosophical Inquiry* – dealing with the value of punishment in the case of a man whose action was necessitated, that is an almost word for word transcription from Hobbes.[28] But it was in their answers to objections to determinism that one can best see the connection between the two. Hobbes gives six objections. Collins divides the first of these – that dealing with the justice of punishment if there is no freewill – into two and his own third objection is an amalgam of the second to the fifth objections given by Hobbes. In his replies he follows the earlier writer. The influence is clear. It remains to be seen how close was the correspondence between his ideas and those of Hobbes.

[24] *Ibid.*, IV, p. 273.
[25] *Ibid.*, I, p. 399.
[26] *Ibid.*, IV, p. 247.
[27] *Ibid.*, IV, p. 273.
[28] *A Philosophical Inquiry*, p. 94.
Hobbes, *English Works*, IV, p. 253.

JOHN LOCKE

John Locke, in the first edition of his *Essay* was a psychic determinist. In this his theory, as it was then formulated, was very similar in many ways to that of Collins. He was, and this was true of all the editions, a hedonist. "Things," he wrote, "are good and evil only with reference to pleasure and pain."[29] "The same thing," he said, "is not good to every man alike."[30] This hedonism was mitigated, or rather contradicted, by other passages in his *Essay* in which he made morality depend upon the rules of right conduct laid down by God.[31] "The will or preference," he wrote, "is determined by something without itself"[32] – namely by good or seeming good. We are necessitated to follow that which we consider to be the greater good. Our wrong choices come because present happiness makes a greater impression on us than that which is future, even though that happiness would be more perfect. Freedom he made to consist, as did Hobbes, in freedom from external coercion. In other words, in talking of freedom he was always discussing the freedom that we might or might not have in carrying out the course of action we prefer.

In the second edition of the *Essay* the chapter on "Power" was considerably rewritten and enlarged. Now, Locke held, our wills are determined by what he described as the greatest uneasiness. By "uneasiness" he meant desire for a good that is absent. This still is determinism. Again, the present uneasiness often has more influence than the consideration of the absence of a good that is of a higher order but more remote. But there is one very important addition and qualification. The mind, said Locke, has "a power to suspend the execution and satisfaction of any of its desires." "In this lies the liberty man has ... in this seems to consist that which is (as I think improperly) called freewill." Man can suspend the execution of his desires – and one should note the emphasis on the idea of the execution of one's desire and not of one's act of will – in order to have a longer time to deliberate as to what is best ... "to examine, view and judge of the good or evil of what we are going to do." This, he said, is a matter of daily experience. But, once we have judged, all we can

---

[29] J. Locke, *An Essay Concerning Human Understanding*, II, 20, n. 2. The edition used is that of A. C. Fraser, republished in Dover Publications, (New York, 1959). This is the only edition which contains all the variant passages of the different editions of the *Essay*. The reference is from I, p. 303.

[30] *Ibid.*, Dover edition, I, pp. 377 and 350. *Essay*, 1st ed., II, 21, n. 34; 2nd ed., II, 21, n. 55.

[31] *Ibid.*, Dover edition, I, p. 364. *Essay* II, 21, n. 72; Dover edition, II, p. 208. *Essay* IV, 3, n. 18.

[32] *Ibid.*, Dover edition, I, p. 375. *Essay*, 1st ed., II, 21, n. 29.

do is "to desire, will and act according to the last result of a fair examination."[33] Now one should note that this is at least postponed determinism. Our freedom at the most consists in suspending the pursuit of that course of action towards which our first desire leads us. When we have deliberated, our judgement dictates our choice. Having decided one course of action is better we have no option but to follow it. This is psychic determinism. But the question remains; was our act of suspension free in a libertarian sense? Anthony Collins criticised this passage in Locke. He seemed to consider that Locke was saying that we are really free, without any prior judgement and debate on the matter, to suspend our action. A. C. Fraser in his edition of the *Essay*, at least doubts whether, even in this choice, Locke leaves man free from "the mechanism of nature" and says that "on his premises, the suspension must be the natural issue of uneasiness."[34] There is no mention in the text itself of uneasiness determining the act of suspension and in a letter of Locke's to Molineux, quoted by Fraser, Locke wrote "though it be unquestionable that there is omnipotence and omniscience in God our maker, yet I cannot make freedom in man consistent with omnipotence or omniscience in God, – though I am as fully persuaded of both as of any truths that I most freely assent to."[35] Now if man were determined by the most pressing uneasiness, as Locke said, and if his choice of suspension were also so determined, then, as Collins held with regard to the Divine foreknowledge, God, knowing man, would know how he would choose, as his choice would be predetermined by his character and natural circumstances, which themselves would be determined by material causation. From internal and external evidence, therefore, a case can be made out for saying that this particular act of choice – of suspension – was really, in Locke's opinion, not necessitated. This would seem to be Collins' view and Collins knew Locke well, had discussed the *Essay* with him and, in Locke's opinion, understood it better than any man he knew. Locke's final position, therefore, might not have been one of complete psychic necessity.

PIERRE BAYLE AND WILLIAM KING

With regard to Bayle, one's approach to the question of freewill must be rather different. Bayle was a Pyrrhonist and fideist. Elizabeth Labrousse, in her life of Bayle, argues very cogently, with regard to this question of

---

[33] *Ibid.*, Dover edition, I, pp. 344, 345. *Essay*, II, 21, n. 48.
[34] *Ibid.*, Dover edition, I, p. 345.
[35] *Ibid.*, p. 316.

freewill, in defence of the genuineness of his fideism.[36] He could well, as she says, have been a man of the Enlightenment, tongue in cheek, posing every difficulty against Christian belief and then blandly and insincerely asserting that one must reject reason and throw oneself blindly into the arms of faith. But it would seem that his fideism was genuine. However, his approach to every problem and in particular to the problems of evil, of grace, of predestination and of freewill – all of which fascinated him – leaves one at something of a loss to decide what his real opinion would have been had he formed it purely on the basis of human reason.

To him the problem of freewill was bound up inextricably with that of grace and predestination. In other words he saw it largely in a theological context. He declared that Thomists, Jansenists and Calvinists alike denied that all the causes which concur with the soul leave her the power of acting or not acting, and that the Molinists asserted that they do.[37] He admitted, of course, that the Thomists would deny they were Jansenists and the Jansenists that they were Calvinists. It would be otiose to try to disentangle Bayle's real thoughts on the whole problem of freewill. Elizabeth Labrousse considers that he ended by fideistically accepting the orthodox Calvinist position and that this implied a rejection of what is known as liberty of indifference. In his *Réponse aux Questions d'un Provincial* he wrote –

De simple Philosophe à Philosophe la dispute de la liberté est au dessus de la décision. Le Molinisme a de l'avantage sur l'autre parti au tribunal de la morale, mais il a du désavantage au tribunal de la métaphysique. Ses principales forces consistent dans les conséquences qui résultent de ce que l'homme agiroit toujours nécessairement: il faut avouer que ces conséquences sont bien terribles, mais un Philosophe qui ne seroit point Chrétien, les afoibliroit beaucoup, soit à cause qu'il n'admettroit pas ce que l'Ecriture nous aprend sur les peines du péché, soit à cause qu'il effaceroit de la liste des péchez un très-grand nombre d'actions que l'Ecriture y renferme.... C'est pourquoi il soûtiendroit plus facilement la définition que Luther & que Calvin ont donnée de la liberté de l'homme.[38]

What he meant was that the Molinists could not reconcile God's foreknowledge and omnipotence with man's freewill but that their opponents would seem to make him the author of sin and to cause a man's damnation without the man having the freedom to avoid the moral evil that led to that damnation. He was elsewhere, however, to argue that the Molinist solution had God place the damned in circumstances in which they could

[36] Cf. E. Labrousse, *Pierre Bayle*, (The Hague, 1964), II, p. 414.

[37] Pierre Bayle, *Historical and Critical Dictionary*, Desmaiseaux's edition, 2nd edition, (London, 1736), III, p. 548, Article "Jansenius," note H.

[38] Pierre Bayle, *Oeuvres Diverses*, (The Hague, 1727), III, p. 782. *Réponse aux Questions d'un Provincial*, II, c. 139.

not avoid their damnation,[39] and that if the Calvinists only examined their own experience they would have a strong persuasion that they were the free cause of their decisions. It was only after consulting their Catechism, he said, that they adopted other principles.[40] But with regard to the proof of freewill from experience he had doubts. Collins paraphrased and translated one of these doubting passages. "They who examine not to the bottom what passes within them, easily persuade themselves, that they are free: but they who have considered with care the foundation and circumstances of their actions, doubt of their freedom, and are even persuaded, that their reason and understanding are slaves that cannot resist the force which carries them along."[41] Bayle himself was to say that this passage "ne peut passer tout au plus que pour une bonne dificulté, ce n'est ni une démonstration, ni une très-forte preuve."[42] In a letter to the Abbé du Bois, of December 13th, 1696, he wrote,

il est certain que notre expérience de liberté n'est pas une bonne raison de croire que nous soions libres; & je n'ai vu encore personne, qui ait prouvé qu'il soit possible qu'un Esprit créé soit la cause efficiente de ses volitions. Toutes les meilleures preuves, qu'on allegue, sont que sans cela l'homme ne pécheroit point, & que Dieu seroit l'Auteur des mauvaises pensées, aussi bien que des bonnes.[43]

And elsewhere he wrote that the revelation of the existence of hell is the only proof of our freedom; or hell would be a punishment, without hope of reformation, of a man who had done wrong by necessity. Bayle, therefore, was a Pyrrhonist, even if he ended in genuine fideistic Calvinism, but the arguments he produced on our experience of freedom and on the problems roused by Divine foreknowledge, and most of all his attitude towards 'liberty of indifference,' had a powerful effect on Anthony Collins.

As these last arguments were directed against William King's theory of freewill, as expressed in his *De Origine Mali*, it will be as well to examine King's position immediately. The Arminians and the Anglicans alike believed in liberty of indifference. It was the keystone of the libertarian position. King gave it his own interpretation. Bayle said that King rejected the traditional theory of indifference and his criticism was directed against King. But his attack developed on a wider front, against the traditional theory as well. Collins, reading Bayle, thought that this theory taught that freewill meant a cold indifference or an indifference to evil

---

[39] *Dictionary*, IV, p. 518; Article "Paulicians," note F.
[40] *Oeuvres*, III, p. 665; *Réponse*, II, c. 83.
[41] *Inquiry*, pp. 27, 28. *Dictionary*, III, p. 374, Article "Helen," note Y.
[42] *Oeuvres*, III, p. 783; *Réponse*, II, c. 139.
[43] *Oeuvres*, IV, p. 726.

as such, so that one could choose evil as evil. This misconception is a cardinal criticism of Collins' book. It would be as well to see exactly what the term meant in its classical interpretation.

LIBERTY OF INDIFFERENCE

St. Thomas Aquinas gives the basis of the doctrine in a passage in his *De Malo*.[44] The will, he says, is only necessitated by that which has a necessary connection with "beatitudo," the "Summum Bonum" of man, that which will make him perfectly happy. But it is clear, he continues, that particular good objects are not so connected, as man can be happy without them. Therefore, when any good object is proposed to a man, he is not necessitated to move towards it. The perfect good, which is God himself, has a necessary connection with man's beatitude, since man cannot be happy without him: but this connection is not clear in this life, since God is not seen face to face ... *Per essentiam* ... and, therefore, man's will in this life does not necessarily cling to God. In the next life, when God is seen face to face, it will be so necessitated. *A fortiori* it would follow that lesser good objects do not here necessitate man's will. In this sense man's will is said to be indifferent to such objects: not that it is coldly aloof, nor that it can choose evil as evil; but that man – and here one has to avoid "faculty language" – realises that he can do without any individual one such object.

Aquinas did not use the term "indifferent." By the seventeenth century it had become current usage. John of St. Thomas (1589–1644) wrote

Quia enim radix libertatis in nobis oritur ex universalitate istius potentiae (i.e. the will) ad omne ens, seu ad omne bonum, inde est, quod quamdiu stat voluntatem operari ex ista universalitate stat operari libertate, quia universalitas illa indifferentiam importat, seu radicem indifferentiae; sed tamen ista indifferentia, et universalitas ita se habent, quod respectu boni limitati, et non adaequantis totam potentiae universalitatem operatur cum indifferentia, et libertate formali; respectu autem boni universalissimi, et summi, sicut est Deus clare visus, adaequatur tota universalitas, et indifferentia voluntatis. Unde erga tale objectum non potest operari indifferenter, licet operetur ex ipsa radice indifferentiae, quae est universalitas voluntatis cum plena advertentia, et haec dicitur libertas eminenter.[45]

In other words the intellectual appetitive being is orientated towards universal good. Only such a completely good object can make him perfectly and contentedly happy. Limited good objects will not satisfy him and he can abstain from choosing them, though if he does choose them

---

[44] Thomas Aquinas, *De Malo*, q. III, a. 3, Corpus.
[45] John of St. Thomas, *Cursus Theologicus*, (Paris, 1885), V, p. 370, *In Primam Secundae D. Thomae*, q. VI, a. II, n. 13.

he must choose them *sub specie boni*. Faced with the universal, the perfect good, the existence of which is the root of his indifference to lesser good objects, he cannot but choose it and this is what John of St. Thomas calls the realisation of freedom "eminenter," as it is the complete and indeed demanding satisfaction of the intellectual appetitive being. This "indifference" was very different from the cold aloofness which Collins visualised. The word itself had misled him and a correct understanding of it would have provided answers to many of his and Bayle's objections to "liberty of indifference," even though they might not have satisfied them. Moreover, the idea of indifference might supply an answer to the problem as to why a man can choose the lesser of two good objects. He is necessitated to choose neither, as both are limited. He can, therefore, choose either, reject either, or reject both. Individually neither compels his choice. And his choice of the lesser good would be reasonable as, being a reasonable appetitive being, he realises, explicitly or implicitly, that neither object, is, absolutely speaking, necessary to his happiness, while his choice would not be uncaused as the active and unnecessitated intellectual appetite is itself a cause. An animal, presumed to be without such self-conscious awareness, would have no such power of "indifference" and no such freedom of choice. This would not solve all the problems connected with freewill, least of all the theological difficulties advanced by Bayle, which were, of course, by no means novel, but it is of significance in dealing with those which he raised concerning the idea of liberty of indifference and with the defence of determinism that was advanced by Anthony Collins.

The idea of indifference is most appropriate to the problem with regard to the choice between alternatives in the rather over-simplified way in which it was at times advanced by Collins. Such choices are rarely simple. It is extremely difficult to decide which is the better of two alternatives if they are of different orders – the choice of going for a walk, for example, presuming that exercise is not urgently necessary, and of listening to a classical concert. Many such choices are really not choices at all but intellectual assessments carried out as a result of a previous decision – to invest a sum of money, for example, in the most promising shares. Habit, education, circumstances play their part (and Collins, like Leibniz, did not ignore this aspect of the situation). But with regard to the rather over-simplified question – how can a rational man choose the lesser of two good objects – the idea of indifference is important. Among the reasons for a choice an "indifferent" intellectual appetite is an important and indeed a final one. One prefers because one prefers, one rejects because one rejects: one knows what one is doing. The man who chooses is not choosing

blindly. The choice is neither blind nor random but deliberate. Will or rational appetite is an ultimate factor in human nature. It can well be argued that the idea of "indifference" is basic to human freedom. Even taken out of its theistic setting it could still be of value. Is any individual limited good object necessary to man?

DE ORIGINE MALI

The most provocative statement in *De Origine Mali* is that "the goodness of the object does not precede the action of election, so as to excite it, but election makes the goodness in the object." It is immediately qualified by the explanation "that is, the thing is agreeable because chosen and not chosen because agreeable."[46] "Good" here is being taken in the sense of "agreeable." The statement can be linked with King's assertion that "it proceeds immediately from his will that things please God, i.e. are good," which again is qualified, though in a rather different way, by the preceding clause "'Tis evident that the Divine will was accompanied in this, as in all other cases, by his goodness and wisdom"[47] and the later statement that "things are made by God with goodness, that is, with a certain congruity to his own nature."[48] There is more than an echo of Descartes' idea, criticised by Ralph Cudworth, that the goodness of things depends on God's will and that God did not will "that the three angles of a triangle should be equal to two right angles, because he knew it could not be otherwise. But on the contrary ... because he would that the three angles of a triangle should necessarily be equal to two right angles, therefore this is true and can be no otherwise."[49] Cudworth was criticising the idea that moral good and evil depend on the arbitrary will of God. What Descartes wanted to do was to safeguard the omnipotence of God. This led him at times into statements that implied that God could create things that would be in fact absurdities, a position which other of his writings contradicted.[50] As can be seen from his qualifications, King did not hold that goodness in things depends on arbitrary will, but he did go a good way towards making will an arbitrary power. Goodness in created things,

[46] William King, *An Essay on the Origin of Evil*, (London, 1732), Edmund Law's translation, II, pp. 279, 280.
[47] *Ibid.*, p. 287.
[48] *Ibid.*, p. 293.
[49] R. Cudworth, *A Treatise concerning Eternal and Immutable Morality*, (London, 1731), p. 29. R. Descartes, *Oeuvres Philosophiques*, Édition de F. Alquié, (Paris, 1967), II, pp. 872, 873.
[50] For a discussion of what is involved in this question and of Descartes' position, cf. A. Boyce Gibson, *The Philosophy of Descartes*, (London, 1932), pp. 270–283.

in Cudworth and, for that matter, in Descartes, had an ontological relationship with God. They must mirror his Essence. So also, in King, goodness has a "certain congruity" with God's nature and is in accord with his wisdom. But the goodness of which he was talking in the sentence quoted at the beginning of this section seems to be goodness taken in an hedonistic sense. As has been said, things are good because they are agreeable, and when our will makes a thing good by choosing it, King means it makes it agreeable. At the same time he was not an hedonist. He admitted absolute standards of right and wrong and of good and evil. He said we can make wrong elections. "We have beheld not a few disregarding the intreaty of their friends, the advice of their relations, the dictates of their own mind, dangers, distresses, death, the wrath of God, and the pains of hell; in short, despising all that is good, or could appear to be so, when set in competition with such things as, exclusive of the goodness which they receive from election, are mere trifles and worth nothing at all; such as have no manner of good or pretence of good in them."[51] King's theory, therefore, is a very complicated one and he is using the word "good" in a very personal way. It is not surprising that his critics found him a little hard to interpret correctly, particularly when, as Leibniz said of Bayle, they had only read his book in reviews and not in the original.

There were other complexities in King's position. He criticised some of the defenders of what he considered to be liberty of indifference.[52] He was thinking of Locke. He expressly mentioned the idea that the will can suspend its act. Now this was not exactly what Locke said. Locke said the will can suspend the execution of its desires. Nor would Locke be considered a leading protagonist of liberty of indifference. Again King did not, on other points, give an accurate account of Locke's theory – if indeed he had Locke solely in mind. Moreover, he considered his own theory to be one of liberty of indifference.

With regard to the influence of the intellect there was also some ambiguity. The intellect, he said, "cannot pronounce upon the goodness or badness of them [objects], till it perceives whether the power [the will] has embraced or rejected them."[53] He went so far as to say that the will has such power over the understanding that it can force it to admit falsities for truths.[54] However, he also said that the understanding tells us what

---

[51] King, *An Essay*, II, p. 340.
[52] *Ibid.*, p. 254 sq.
[53] *Ibid.*, p. 273.
[54] *Ibid.*, p. 341.

is possible, so that we might avoid the frustration of willing what is impossible and that it is used "as a light before us to distinguish good from evil" before the act of election, though it is used" as a judge and a counsellor, not as a sovereign and dictator."[55] It "points out to us and admonishes us ... to avoid these external evils or to embrace the good; but till we have exerted an act of election about them, these do neither become absolutely agreeable, nor the others odious."[56] The word "absolutely" certainly implies some modification of his earlier statements. The fact would seem to be that King was so anxious to assert the freedom of the will, so intent on preserving it from determination by the judgement, that he made it a more arbitrary power than would the more typical defenders of freedom of indifference. If, as he says, the goodness of the object does not precede nor excite the act of election, it would seem logically that it is not chosen *sub specie boni*. On the other hand, when he modifies this by saying that it is not absolutely good and that the understanding is a counsellor admonishing us to avoid evil and embrace good, he comes much nearer to the position of the orthodox defenders of liberty of indifference. His language is not consistent. It left him very open to attack and, at the same time, because of this inconsistency, an attack on his more arbitrary statements could be taken as criticism of the very different theory that had long gone under the title of that of liberty of indifference. There is, however, one factor in the exercise of choice towards which King's theory might point. The will, or rather the intellectual appetitive being, by deliberately concentrating on one alternative and deliberately relegating another to the periphery of one's consciousness, can increase the attraction of the first object. Again this does not solve the problem of free will, but it may provide a partial elucidation of the mechanism of choice, particularly in the type of choice that involves a moral struggle, concerning an alternative that is unethical but profitable or attractive.

BAYLE AND KING

Pierre Bayle did not read King's work itself. He only knew of it through a digest given by Jacques Bernard in his *Nouvelles de la République des Lettres* of 1703.[57] This did not prevent him writing a long critique of King's theory, printed in his *Réponse aux Questions d'un Provincial*. He recognised that King did not approve of those who make freewill consist

---

[55] *Ibid.*, p. 361.
[56] *Ibid.*, p. 362.
[57] Bayle, *Oeuvres*, III, p. 650; *Réponse*, II, c. 74.

in liberty of indifference.[58] At the same time many of his criticisms were directed at the latter as well as at King himself. He argued from experience, saying that in order to be pleased with one's choice one need not consider oneself free from external causation.[59] Those, for example, who consider they have been guided in their decisions by God are all the more pleased by this consideration. Those who receive an unexpected windfall – as does a peasant who finds by accident a treasure hidden in his garden[60] – are all the more pleased because their gain was not the result of premeditated choice. Collins gave the same argument with almost the same example – that of finding a treasure on the road. King, in his more radical statements, those that seem to imply that all the appetibility of an object comes from its being freely chosen, left himself open to this attack: but he also implied that the intrinsic appetibility or distastefulness of a thing affected the appreciation of those who chose it, at least to the extent that an unpleasant object is less pleasurable to a man, even though his will can make him to some extent happy in choosing it.[61] Men, said Bayle, would not be at all happy at being given a power which made them indifferent to "beatitude." In a footnote he added that both experience and the opinions of all philosophers tell us that we cannot choose evil as evil.[62] Man, he said, is not indifferent to good as such, and therefore it cannot be a defect in man to be determined with regard to particular good objects. Indeed, he said, some philosophers hold we are so determined.[63] The blessed in heaven have no such freedom of indifference[64] and if man had the sort of freedom which King allows him he would indeed be ungovernable.

S'il a une liberté indépendante de la raison, & de la qualité des objets clairement connuë, il sera le plus indisciplinable de tous les animaux, & l'on ne pourra jamais s'assurer de lui faire prendre le bon parti. Tous les conseils, & tous les raisonnemens du monde pourront être très-inutiles; vous lui éclairerez, vous lui convaincrez l'esprit, & néanmoins sa volonté fera le fiere, & demeurera immobile comme un rocher.

Collins reproduced this last passage in translation, almost word for word, though he did not refer to Bayle:[65] he also used the argument of the state of the blessed in heaven, but this line of attack was a common one in the

---

[58] *Oeuvres*, III, p. 658; *Réponse*, II, c. 80.
[59] *Oeuvres*, III, p. 658 sq.
[60] *Ibid.*, p. 661; *Inquiry*, p. 72.
[61] King, *An Essay*, pp. 364, 365.
[62] *Oeuvres*, III, p. 662; *Réponse*, II, c. 81.
[63] *Oeuvres*, III, p. 679; *Réponse*, II, c. 90.
[64] *Oeuvres*, III, p. 676; *Réponse*, II, c. 89.
[65] *Oeuvres*, III, p. 679; *Réponse*, II, c. 90. *Inquiry*, pp. 79, 80.

defence of determinism. Now King never said that man can choose evil as evil, though he did declare that the will can make the understanding take evil things for good.[66] Bayle declared further that King should not reject the Scholastic doctrine of indifference:

car ils – the Scholastics – la délivrent de toute sorte de détermination qui ne dépend pas de la volonté; ils disent que la volonté peut aimer; ou n'aimer pas, ou haïr même une chose, quelles que soient les dispositions des autres facultez de l'ame: ils disent qu'elle n'est point necessitée par le jugement pratique de l'entendement; & si quelques-uns d'entre eux avouënt qu'elle se conforme toûjours à ce jugement pratique, ils prétendent qu'elle a influé sur ce jugement, & qu'elle le peut écarter, & en faire substituer un autre.[67]

In spite of his earlier statement, therefore, he was approximating the traditional doctrine of indifference to that of King; and Collins followed his example. Now King certainly was, to say the least, ambiguous. At times, as has been said, he seemed to be approaching the position of the Scholastic philosophers. But to confuse their theories with the more extreme statements of King was an error on the part of Bayle. They did not so separate the will from the rest of man. They did not make it a completely arbitrary power. They did not, as a result of their doctrine of indifference, make man impervious to arguments. On the contrary they raised the objection against determinism, that it would make argument useless. As has been said, the root of "indifference" lay, not in an aloof coldness, but in the fact that an intellectual appetitive being realises that no limited good is necessary to its "beatitude." It cannot choose evil as evil, but it can choose a lesser good *sub specie boni*. It can choose an object that is, taken as a whole, morally evil, because that object will also possess a degree of goodness. It can concentrate its attention on that goodness and ignore or push to the periphery of its consciousness the evil which it contains, if one can speak of evil as a positive thing and not as a deficiency. It can decline to look at the ethically good alternative. Bayle's error in misinterpreting the meaning of indifference may have been, in part, due to King's setting up will and intellect almost as antagonists and making will, at times, appear an almost arbitrary power.[68] Whether

---

[66] King, *Essay*, p. 341.

[67] Bayle, *Oeuvres*, III, p. 677; *Réponse*, II, c. 89.

[68] Bayle was not always consistent in his statements on indifference. He criticised Yves de Vallone, ex-Canon of St. Geneviève in Paris and convert to Calvinism, for misunderstanding the theory. He wrote, "Ils – the Scholastics – n'ont jamais prétendu qu'elle consiste dans l'indolence, ou qu'elle excluë les inclinations, et les plaisirs prévenans. Ils n'entendent autre chose par cette espece de liberté qu'une puissance de se déterminer soi-même, et de résister si l'on veut aux impressions des objets et aux passions etc." (*Oeuvres*, III, p. 856; *Réponse*, II, c. 168). This is not quite in accord with his statement, in his criticism of King, quoted above, that the Scholastics say "la volonté

this be so or not, his misinterpretation had a considerable effect on Collins' argument in his *Philosophical Inquiry*.

LEIBNIZ

The last writer who needs to be considered is Gottfried Wilhelm Leibniz. Although Collins' basic principles appeared in his earliest works, before the publication of Leibniz's *Theodicy*, he certainly read and quoted the latter book. Moreover the *Theodicy* was a very courteous polemic against Bayle. It dealt with Bayle's opinions on the problems of evil and of free will and had appended to it critiques of both Hobbes' controversy with Bramhall and King's *De Origine Mali*. In many ways Collins' ideas resembled those of Leibniz and to some extent his opinions were influenced or at least supported or modified by the writings of the latter philosopher.

Leibniz's theory of freewill was founded on what he termed the principle of determinant reason, by which he declared "that nothing ever comes to pass without there being a cause or at least a reason determining it, that is, something to give an *a priori* reason why it is existent rather than nonexistent and in this wise rather than any other."[69] This would hardly be questioned by the libertarians, but it was Leibniz's application of the principle that was open to dispute. It was from it that he concluded that the world which God created is the best of all compossible worlds and that man in all his acts – including those in which his will takes a part – is determined to one course of action either by his judgement of what seems best or, more accurately, by his conscious judgement and the unconscious influence of circumstances upon him. "The will is never prompted to action save by the representation of the good, which prevails over the opposite representations." "There is always a prevailing reason which prompts the will to its choice" – and by "prompts" he means "determines" – but, he adds, "and for the maintenance of freedom for the will it suffices that this reason should incline without necessitating."[70] This sounds like libertarianism, but to understand what he means by "necessity" one has to have recourse to his second great principle, that of "contradiction, stating that, of two contradictory propositions, the one is true the other false." An act, according to him, can be "determined" and "contingent" without being "necessitated."[71] In other words,

peut aimer, ou n'aimer pas, ou haïr même une chose, quelles que soient les dispositions des autres facultez de l'ame." It was Bayle's remarks on King that influenced Collins.

[69] G. W. Leibniz, *Theodicy*, translated by E. M. Huggard, (London, 1951), p. 147.
[70] *Ibid.*, p. 148.
[71] *Ibid.*, p. 147. Note the meanings Leibniz attaches to the words "necessitated"nd a "determined."

although the act of will is determined by reason and circumstance, it is not a logical contradiction that it should not have been posited. In the same way it was not a logical contradiction or, to use Leibniz's term, a "metaphysical impossibility" that God should have created another world. Leibniz's position with regard to freewill is summed up in three sentences in his critique of King.

As for me, I do not require the will always to follow the judgement of the understanding because I distinguish this judgement from the motives that spring from insensible perceptions and inclinations. But I hold that the will always follows the most advantageous representation, whether distinct or confused, of the good or the evil resulting, from reasons, passions and inclinations, although it may also find motives for suspending its judgement. But it is always upon motives that it acts.[72]

These motives he regards as factors, other than the will, which in every case determine it. The word "motive" is a deceptive one. As Clarke was to reply to Collins; "when we say, in vulgar speech, that motives or reasons determine a man, 'tis nothing but a mere figure or metaphor. 'Tis the man that freely determines himself to act."[73] But Leibniz made his point clear. "All is therefore certain and determined beforehand in man," – the word "beforehand" is important – "as everywhere else, and the human soul is a kind of spiritual automaton, though contingent actions in general and free action in particular are not on that account necessary with an absolute necessity, which would be truly incompatible with contingency."[74]

He rejected King's idea of indifference.[75] But he also rejected the traditional idea of liberty of indifference as well, while keeping the word "indifferent." "There is therefore a freedom of contingency or, in a way, of indifference, provided that by "indifference" is understood that nothing necessitates" – one should hold in mind his meaning of the word "necessitates" – "us to one course or the other; but there is never an indifference of equipoise, that is where all is completely even on both sides, without any inclination towards either. Innumerable great and small movements, internal and external, cooperate with us, for the most part unperceived by us ... it suffices that the creature be predetermined by its preceding state, which inclines it to one course more than to the other."[76] Collins was to toy with this idea of equipoise, having in mind no doubt, both

---

[72] Leibniz, *Observations on the book concerning "The Origin of Evil,"* published with *Theodicy*, p. 418.

[73] Clarke, *Remarks*, p. 11.

[74] *Theodicy*, p. 151.

[75] *Theodicy, Observations*, p. 416.

[76] *Theodicy*, pp. 148, 149.

Leibniz's remarks and Bayle's article on Buridan. Why can there be no such equipoise? Leibniz deduced the idea as a subsidiary principle from that of sufficient reason. God could have no reason for creating two beings that were identical, as they should occupy exactly the same position in the scale of creation and therefore be not two but one. And in practice, he said, equipoise is impossible. Dealing with Bayle's article, he said, "the case of Buridan's ass between two meadows, impelled equally towards both of them," – and therefore unable to choose and starving in the midst of plenty – "is a fiction that cannot occur in the universe ... There will always be many things in the ass and outside the ass, although they be not apparent to us, which will determine him to go to one side rather than the other."[77] He rejected Bayle's suggestion that the dilemma might be solved in man's case by "'the pleasing fancy that he is master in his own house, and that he does not depend upon objects.' This way is blocked for ... that has no determining effect, nor does it favour one course more than the other."[78] But he did approve of Bayle's remarks in which he approximated King's idea of indifference to that of the Scholastic. He quoted Bayle at length and gave his reason; "a freedom of indifference, undefined and without any determining reason, would be as harmful and objectionable, as it is impracticable and chimerical."[79] It would be, he said, irrational. Here Leibniz would strengthen Collins' misinterpretation of the meaning of liberty of indifference. Of Hobbes, Leibniz wrote with remarkable tolerance and he agreed with Hobbes' answers to many of the objections he himself, i.e. Hobbes, raised against freewill,[80] but he declared that both Hobbes and Spinoza went too far "destroying freedom and contingency: for they think that that which happens is alone possible, and must happen by a brute geometrical necessity."[81] Curiously, Collins, in a single passage, takes up the position of a logical determinist and here follows what Leibniz considers to be the position of Hobbes and Spinoza rather than that of Leibniz himself and Bayle. Collins wrote: "It was as impossible for Julius Caesar not to have died in the Senate, as it is impossible for two and two to make six."[82] Clearly he had in mind Leibniz's statement, quoting Bayle; "It is today a great embarrassment for the Spinozists to see that, according to their hypothesis, it was as impossible from all eternity that Spinoza, for instance, should not die at The Hague,

---

[77] *Ibid.*, p. 150.
[78] *Ibid.*, p. 311.
[79] *Ibid.*, p. 316.
[80] *Ibid.*, p. 160.
[81] *Ibid.*, p. 348.
[82] *Inquiry*, p. 107.

as it is impossible for two and two to make six." Leibniz regarded Spinoza's death at The Hague as a contingent though "determined" event, not contrary to the principle of contradiction. He criticised Bayle for spoiling his remarks by adding "Now what contradiction would there have been if Spinoza had died at Leyden? Would nature then have been less perfect, less wise, less powerful?" He said that Bayle "confuses here what is not possible because it implies contradiction with what cannot happen because it is not meet to be chosen."[83] But Collins seemed to be differing deliberately from both Bayle and Leibniz. What will be of interest will be to see how the ideas of Hobbes, Locke, Bayle and Leibniz are interwoven in his thought and how far his own work bore the stamp of an original thinker, how far, in fact, he deserved the accolade given him by Voltaire and Priestley.

THE *INQUIRY*

The *Inquiry* opened with a Preface in which Collins clearly stated his position in the debate: he asserted he was a defender of liberty, but of a liberty that is compatible with what he called moral necessity: namely determination by man's reason and senses. He claimed that the majority of philosophers, "however opposite in words they appear to one another," agreed with him. Among these philosophers he numbered Locke and, from his express rejection of "absolute, physical, or mechanical necessity," it would seem he was dissociating himself from Hobbes. It is clear he intended to take the war into his opponents' territory. Their arguments had often been that necessity destroys religion and morality. He, on the contrary, asserted that it was his opponents whose theories were subversive of both and that he would show this. It was a clear statement of the line his argument was to take.

His *Introduction* is, one is perhaps inclined to think, a rather long epistemological digression based on Locke's *Essay*. It shows his insistence on clear and distinct or, the terms Locke preferred, "distinct and determinate" ideas. He asserted, a little unnecessarily in this context, that men can have clear and distinct, if inadequate, ideas of God. There is an echo here of his *Essay on the Use of Reason* and a brief, but undeveloped, reference to the Trinity. But he made no attempt to impugn again the idea of mystery. His thoughts here were of more interest when taken in conjunction with his *Vindication of the Divine Attributes*. But the point he wanted to make was that many of his opponents had not had clear

---

[83] *Theodicy*, p. 235, on Bayle, *Dictionary*, II, p. 496, article "Chrysippus," note S.

and distinct ideas of what they meant by freedom. He had no doubts
about his own capacity. The argument was to be of significance when he
came to deal with the definitions of freedom given by different philosophers.

He then gave his own definition:

Man is a necessary agent, if all his actions are so determin'd by the causes preceding
each action, that not one past action could possibly not have come to pass, or have been
otherwise than it hath been; nor one future action can possibly not come to pass, or
be otherwise than it shall be. He is a free agent, if he is able, at any time under the
circumstances and causes he then is, to do different things: or, in other words, if he is
not unavoidably determin'd in every point of time by the circumstances he is in, and
causes he is under, to do that one thing he does, and not possibly to do any other.

It is a good definition of liberty and necessity but it leaves open the question as to what kind of necessity determines man. It is as open to a
Hobbesian interpretation as to that favoured, in his Preface, by Collins.
Leibniz would, on the whole, agree with Collins' first statement but he
did not agree with Hobbes. He would say that the factor of intellect ex-
cluded mechanical fatality. Many of the defenders of freewill would say
that intellect demands freedom of indifference. And Hobbes allowed for
intellect. It can be argued that only two alternatives present themselves,
liberty of indifference or complete causal determinism, in which intellect,
if it is to be found, is included as a cog in a causally determined machine.

Collins then went on to deal with the argument from experience. He
went straight to the heart of the question; the conviction that the average
man has that he is free and free with what one is inclined to think, though he
would not recognise it under the antique terminology, is liberty of indifference. Collins gave an alternative explanation of the conviction of
freedom. Men, he said, do not notice the causes which influence them.
Their feelings of guilt represent only a return to a more ordered existence,
with a feeling of regret for past disorder, though both order and disorder
were causally necessary. They, he said, feeling free from the past causality,
believe they could have chosen otherwise. They also notice that at different
times – though under different circumstances to which they do not properly advert – they choose differently and think, therefore, that they
choose freely. The points he made were not new. Hobbes when he wrote
of the spinning top that thought it spun as it willed, if it were not aware
of the boy's whips, Bayle's weathercock and Leibniz's minor unnoticed
causes influencing men's actions add up to much of what Collins was
saying.[84] But rarely has the case for an alternative explanation for what

[84] Hobbes, *English Works*, V, p. 55. Leibniz, *Observations*, in *Theodicy*, p. 432.
Bayle, *Oeuvres*, III, p. 786; *Réponse*, II, c. 140.

men consider to be the experience of freedom been put so clearly and so succinctly. One can see why Collins influenced a man like Voltaire. The libertarian would reject the alternative and would deny that it really explains the feeling of culpability. But Collins made a very plausible case.

Then he came to definitions of liberty. His point was to show that the opponents of necessity, in their definitions of freewill, really agreed with his opinion. He did not succeed in every case. What he did was to show, even if unconsciously, what he had insisted on at the start – the need for clarity of definition. He quoted Cicero. Here he was in error. Cicero was not talking of freewill but of the external freedom such as that enjoyed by princes.[85] La Placette and Jaquelot, both defenders of freewill, gave definitions that could be accepted by a determinist of Collins' school. He could say to Jaquelot that when he wrote "we should even do the contrary if we willed it" that this was true, but that our will was due to necessary causes. Locke's definition certainly fits in with Collins' opinion, but we have seen the difficulties in which Locke was involved. Later Collins was to advert to the modification made in the second edition of Locke's *Essay* to the effect that freedom consists in man being able to suspend the execution of his desires, a theory that he was to reject. All the definitions, said Collins, can be interpreted as declaring that freedom consists in "freedom from outward impediments of action." And he was quite correct. But that did not mean that Jaquelot and La Placette were determinists. Their definitions were inadequate. One needs to read more of their works to see their real position.

There follow three other definitions of freewill: those of Alexander of Aphrodisias, of Bishop Bramhall and of the Arminian Jean Leclerc. Alexander was chosen, presumably, because in Collins' words he was "esteemed the best Defender and Interpreter" of Aristotle. Certainly in the passage cited by Collins and in that which follows it in Alexander's *De Fato* he is shown to be a psychic determinist, in the sense that he held that our acts necessarily follow the judgement of our reason. This was Aristotle's position in his *Nicomachean Ethics*, though Aristotle fell into some difficulty in explaining how a man can perform a wrong act, knowing it not to be to his greater good.[86] But Bramhall was very badly treated. His opinion was grossly misrepresented by Collins, either by selective quotation or by a careless misinterpretation of his words. Collins quoted him as saying that "motives determine not naturally but morally" and asserted that the Bishop was putting forward the same "moral determinism" as he himself. Bramhall, however, had explained his meaning.

[85] Cicero, *De Officiis*, I, 20, 70.
[86] Aristotle, *Nicomachean Ethics*, VII, 3, 5 sq.

"The will is moved by the understanding, not as by an efficient having a causal influence into the effect, but only by proposing and representing the object."[87] Again Collins gave him as saying "Admitting that the will follows necessarily the last judgement of the understanding" and that this this is a "hypothetical necessity." By this and the previous quotation, he said Bramhall was admitting that "choosing or refusing is morally and hypothetically determined." Actually what Bramhall was doing was aiming at showing that "supposing, but not granting" – and these last words were omitted by Collins – that the will did follow the last dictate of the understanding – Aristotle's position in fact – it would still not be subject to the absolute necessity that Hobbes was defending. Bramhall's position is very different from that of Collins.

With regard to Leclerc, Collins gives a rather curious argument. Leclerc was reviewing Locke's *Essay* and, in particular, the chapter on "Power," in the *Bibliothèque Choisie*. He said that the mind is in a state of indifference until the moment when it acts: then, of course, it is determined, as it cannot be performing an act and not performing it at the same time: but it remains indifferent as to the continuation of the act. Collins said that if man really were indifferent he should have it in his power not to be indifferent at the same time. This really is a quibble. Leclerc could reply that if indifference towards a particular course of action is referred to, the mind could come out of it at any time it chose, but that with regard to indifference as a power of the mind, that this is a constant quality which it cannot lose. In his use of these last three sets of quotations, therefore, Collins does not come out with a great deal of credit.

He was more successful in the next part of his argument. He used Erasmus, Leclerc, Gilbert Burnet, Ochino and King to show that all these authors considered freewill to be a very difficult question. "But how can all this happen in a plain matter of fact, supposed to be experienced by everybody? ... This could not happen if matter of fact was clear for liberty." This is a fair statement. Freewill, that is taken for granted by the man in the street, is open to many difficulties. But the men he quoted – and all except Ochino defended freewill, while Ochino himself was attacking predestination – thoroughly realised this. A facile assumption that what we seem to experience must be so should be questioned. But on the other hand there are many other matters of every day experience that are taken for granted, that are open to difficulty and that nevertheless can correctly be taken to be matters of fact, even though they cannot easily be explained. Clarke instanced the existence of the outside world.[88]

---

[87] Given in Hobbes, *English Works*, V, pp. 73–75.
[88] Clarke, *Remarks*, p. 20.

No one in fact, he said, takes up a position of solipsism. Collins showed that apparent experience does not settle the question without further critical enquiry. He did not show that because enquiry is necessary the experience is false.

He was a man who was rather fond of airing his erudition as much by citation as by argument. He quoted the Arminian Episcopius – who wrote a treatise on freewill – to show that he only defended freewill because he thought its rejection would be disastrous to religion and morality, and he returned to Alexander of Aphrodisias to argue, on this authority, that all men believe in Fate. Now Episcopius did,[89] as Collins said he did, say that the argument that the will is determined by the understanding had been one that had baffled many defenders of freewill. He did not say that this was the common opinion of the Schools. He did say that its rejection would be dangerous to religion and morals: but he did not say that this was the reason why he was defending freewill. He said it was the reason why he was rejecting the theology of the Schools, whose faculty language had made the argument for intellectual determinism unanswerable. He, like Leclerc, in the article just quoted, said it was much better to talk of the soul or of man. The "will" then, he said, could not be regarded as something that was blind. It is the knowing intelligent being who chooses. This is a very useful position to take up. It was not noticed or at any rate not mentioned by Collins.

Moreover, again, with regard to Alexander there came a misuse of a quotation. Alexander did not say that all men believe in fate in the sense that all men are fatal determinists. His words were "Esse igitur aliquid fatum, causamque illud censendum esse, quod quaedam fataliter fiant, argumentum satis firmum exhibet anticipata hominum opinio, aut persuasio."[90] But Bayle and Leibniz, in the passages given by Collins, support his point of view in so far as they say that men are often acted upon by subconscious pressures and are – or the Pyrrhonist Bayle would say, may be – acting under necessity when they think they are acting freely.

This concluded the first part of Collins' argument. What had he proved? He implied that he had shown that "some give the name liberty to actions which when described, are plainly actions that are necessary": that others contradict their appeal to an argument from "vulgar experience" by treating it as a very intricate matter: that others defend freewill for ulterior reasons which have no direct bearing on the argument itself: but that "the most discerning" see that it cannot be proved by the argument

[89] M. S. Episcopius, *Opera Theologica*, (Amsterdam, 1650), part II, pp. 198–200.
[90] Alexander Aphrodisiensis, *De Fato*, (London, 1658), p. 10.

from experience: and that "the bulk of mankind have always been convinced that they are necessary agents." In fact he had shown that definitions of freewill are often inadequate: that its defenders admit that there are considerable philosophical difficulties attending it – and none of them with any pretensions to philosophy, would deny this. He had given one man, Episcopius, who he, inaccurately, said was only defending freewill because he thought determinism a threat to religion and morality: and he had found a major philosopher, Leibniz, who agreed with his theory. His assertion that the bulk of mankind believe in necessity, like his earlier statement that without it morals and religion were in danger, was an aggressive foray into the enemies' camp. He had given many quotations, too many of which were inaccurate or misleading. His contemporary readers might not bother to examine the sources for themselves. But his main achievement had been to set the cat among the pigeons. He had given a plausible alternative explanation for the common feeling of freedom. He had shown some confusion in his opponents' ranks. He had, in fact, as he said "pav'd the way, by shewing that liberty is not a plain matter of experience." Those who were not prejudiced by his carelessness or tendentiousness in quotation would have to admit, if they had taken liberty of the will for granted before, that here was an open question that demanded a critical enquiry. This was his main purpose and to this enquiry he was now to proceed.

He began it, rather unnecessarily one would think, by proving that perception and judging of propositions – that they are self-evident, evident from proof, probable, improbable, doubtful or false – are actions in which the mind is under necessity. Perhaps he did this because he was later to accuse Leclerc of saying that we are free to pass different judgements on the same propositions at the same time. Perhaps he did it to show his thoroughness. After this preliminary canter he passed on to the main issue – the question as to whether will is free or not.

He began by defining will. Here perhaps there is a certain tendency to confuse the words "to prefer" with those of "to will." It is only a minor carelessness at the worst. He asked "Whether we are at liberty to will, or not to will" and answered correctly that, when faced with some choice, we must have some will about it, even if it be to defer willing. Then he came to the suggestion of Locke in the second edition of his *Essay*. He said that Locke said that men are free to defer willing. The point has been discussed. Actually, what Locke said was that men can suspend the satisfaction of their desires and that in some cases, in order to make a sound decision, they have an obligation to do so. Collins remarked that this act of suspending willing would not be a privileged case. There would

need to be a reason for it and therefore it would not be unique. If ordinary acts of will are not free neither would this be. He has a very good point, even though it was the suspension of the satisfaction of our desires about which Locke was speaking. Moreover, Locke did give a reason for suspension – to make sure we are following the correct course – or at least he gave a reason why we have an obligation to suspend our desires. As he put it, his addition to the first edition of his *Essay* did not save freewill. But there is a different way of looking at suspension of willing, not that of Locke, but that which is implied by the theory of liberty of indifference. By this in every act of will there is the possibility of suspense, or simply of not choosing an object, as an essential part of the choice. It is not something distinct from a choice or a special kind of willing.

Collins dealt with choices between alternatives. Here he said "without much dispute" we are necessitated to choose what seems the better course. He was following Locke in the first edition of his *Essay*. But he made a major error which, said Clarke, runs through all his work. He confused judging and willing. "Willing or preferring, is the same with respect to good and evil, that judging is with respect to truth or falsehood." It is true that there is a real analogy here. But, he went on, "It" ... i.e. willing ... "is judging, that one thing upon the whole is better than another, or not so bad as another." This is a bad mistake, and by identifying judgement and will it would inevitably make judgement dictate to will. He went on "wherefore as we judge of truth and falsehood according to appearances, so we must will or prefer as things seem to us, unless we can lye to ourselves and think that to be worst, which we think best." The difficulties of comparing alternatives have been discussed earlier. Collins, however, had put forward one of the strongest arguments for determinism, in spite of his major error. He went on to quote Locke, Norris, Bayle and Plato as supporting his case. The passage from Locke was ill-chosen. In it Locke was not discussing this particular problem. Norris, the last of the Cambridge Platonists, though an Oxford man, did think that "the soul necessarily wills as she judges." He explained freewill by "the indifferency of the soul as to attending or not attending, or attending more or less" to the different objects placed before it. He was taken to task for his opinions by Henry More who regarded his theory on attention as "an invention ingeniously excogitated" to escape from the difficulty into which he considered Norris had fallen by his earlier remarks. Bayle, in the passage quoted, compared the soul to a balance, in which the heavier motive moves the will.[91] But Bayle's Pyrrhonist position has also

---

[91] John Norris, in the *Letters Philosophical and Moral to Dr. Henry More*, printed with Norris' *The Theory and Regulation of Love*, (Oxford, 1688), in the *Appendix* to the

been discussed. Plato can be interpreted as being an intellectual determinist, holding sin to be error. On the whole therefore, Collins' authorities support him, though the assistance of authorities does not prove a case. But when he tried to prove his point by using the opinion of "the greatest modern advocates for liberty" – Bramhall was his authority – in their saying "that whatever the will chuseth it chuseth under the notion of good," and "the object of the will is good in general, which is the end of all human actions," he was off the point. Both statements were accepted by the defenders of liberty of indifference, but they did not involve the necessary choice of the greater seeming good. If one chooses a thing, one chooses it *sub specie boni*; but one is not obliged to choose it at all.

It was, therefore, the theory of liberty of indifference that Collins really had to destroy. He did not, in this part of his book, do so very effectively, as he plainly did not understand it. He here took it in one sense to mean a freedom to choose between objects which are indifferent because they are identical. This, of course, was not what was meant by "indifference." If freedom is confined to such, he said, it would not amount to much, as identical objects, at best, would be rare and trivial and would not be found in moral issues. Like Leibniz, however, he denied that there would be in practice absolutely identical circumstances, even though there seem to be such to our conscious minds. Even if you had two identical eggs – the example he gave – then your will to eat and your habitual practice or "some particular circumstances at that time" would make you choose one rather than the other. "It is therefore contrary to experience, to suppose any choice can be made under an equality of circumstances."

This was indeed a travesty of the theory of indifference as is Collins' more general interpretation of indifference as meaning a cold aloofness. Moreover in Leclerc's review of Locke's *Essay* a fairly accurate account of the theory had been given, at least to the extent that Leclerc said it meant that man was moved by good in general but not necessitated by particular good objects. He did not go into the deeper philosophical reasons as did John of St. Thomas or Aquinas. But Collins erred in good company. Leibniz held much the same view of the theory. Locke, as Leclerc complained, did not understand it. Leibniz thought it involved an equipoise in objects and circumstances. Collins was to return to the question later in his *Inquiry*.

There is one further point of interest. The influence of Locke, Bayle and Leibniz can be seen. But in dealing with indifference Collins uses words that are reminiscent of Hobbes. He talks of "trains of causes"

Letters, by Norris, n.n. 13, 17; More's third letter, *ibid.*, p. 181 sq; Bayle, *Oeuvres*, III, p. 784; *Réponse*, c. 139.

and, later, uses the axiom "whatever has a beginning must have a cause." This latter is implied in Locke's *Essay* when he wrote "nothing can no more produce any real being, than it can be equal to two right angles,"[92] and he had the actual axiom itself in his first letter to Stillingfleet. But it occurs in Hobbes' *Of Liberty and Necessity*.[93] The point, at this juncture, is worth noting.

What, then, had Collins done? He had made mistakes, in confusing judgement and will, in his use of Bramhall and in his interpretation of indifference. But he had, despite these errors, put the argument for necessity in one of the most persuasive ways in which it can be developed – that the rational man must choose what is seemingly best and that this is what, in fact, he does, even though he may be unaware of the subconscious pressures exercised on him. And, as Voltaire said, he had written a most cogent exposition of the theory, in Voltaire's opinion the most clear and cogent. The strongest opposing theory he had mis-stated. But for that he had some excuse. And King, to whom he was to turn later, had apparently done a good deal to show the theory of indifference in a dubious light.

Collins continued by saying that our actions subsequent to willing are necessitated, unless we are externally impeded. There would be no argument about this. But he continued, by way of confirming his opinion, by comparing the actions of animals and men. It was something of an *argumentum ad hominem*. His opponents all said that the actions of animals are necessitated. But, he argued, they give as much sign of deliberation and choice as is to be seen in human acts. He went into the matter in some detail. Hobbes had used the same arguments and it appears in Bayle's article on Rorarius.[94] The fact is that many actions of men are instinctive. Libertarians could reply to Collins that such actions are not really deliberate and therefore not free, though man may be able to reflect on them in a way in which an animal cannot. But there is an interesting point. Collins said that human thought is of the same kind as animal thought. In his *Vindication of the Divine Attributes* he was to say that it is of the same kind as God's. At the same time he did allow to man notions of honour and virtue. These are abstract ideas. Are they of the same kind as the other types of thought which he here attributed to brutes?

He made one very good point, against Bramhall. The Bishop had said that children are necessitated. When, asked Collins, do they cease to be so? As freewill is supposed to be linked to intellect, he could have asked,

[92] *Essay*, IV, 10, n. 3.
[93] *English Works*, IV, pp. 274, 276.
[94] *Ibid.*, p. 244; Bayle, *Dictionary*, IV, p. 908, Article, "Rorarius," note F.

when, on Bramhall's premises, do we suppose they obtain that intellect? The way he put the question was to say "What different experience have they when they are suppos'd to be free agents from what they had while necessary agents?" It was a very good question. The Scholastics, of course, would assert that children always have a spiritual soul and therefore always have intellect. Their conclusion would be that they always have freewill, though lack of experience might circumscribe its use.

The argument from causality which was next produced was paralleled in Hobbes' *Liberty and Necessity*. It also owed something to Locke's argument for the existence of God, in his *Essay*.[95] A cause, says Collins, must be a necessary cause, otherwise it is not suited to its effects. However, a free cause, presuming the will is such, would be a determinate cause of a volition – insofar as one can regard the volition as distinct from the man choosing, because it would have determined itself to this particular volition. It would, however, be self-determined or free. There would not be an absence of cause, nor of determining cause.

Collins proceeded to another of his forays into enemy territory. Determinism had often been linked with atheism. He tried to turn the tables on his opponents by saying that, on the contrary, freewill was atheistic. He produced the argument from order for the existence of God, associated freewill with the "Epicurean System of Chance" – Leibniz had also associated indifference with Epicureanism – and said that the Epicureans and Sadducees were believers in freewill and atheistic, while the Stoics, Pharisees and Essenes, who were determinists, were regarded as the most religious groups among antiquity. The Pharisees he associated with St. Paul and, on Dodwell's authority, their alleged determinism with his theology. He had grounds for his assertion with regard to the Stoics and Epicureans. His authority for the Jewish sects was Josephus, who, in the passages referred to, was trying, for reasons of policy, to represent the various groups among the Jews as corresponding to the divisions that existed among the Greek philosophers, and was using terms taken from Greek philosophy that were, in fact, not applicable to Jewish thought. The texts, moreover, are open to various interpretations. Certainly the Pharisees did not completely reject freewill, and the Sadducees, politicians though they were, were rather roughly treated.[96] But Collins' point was to turn his opponents' arguments against themselves.

He continued the same policy by asserting that liberty is an imperfection and necessity a perfection. He took three alleged definitions of liberty,

---

[95] Hobbes, *English Works*, IV, pp. 274, 276; Locke, *Essay*, IV, 10, n. 3 sq.

[96] Cf. G. F. Moore, *Judaism in the first Centuries of the Christian Era*, (Cambridge, Mass., 1927), I, pp. 456–458.

those given, he said, by Leclerc, King and Cheyne. Liberty, he at least implied, was said by Leclerc to be a power to pass different judgements at the same instant of time upon propositions that are not evident. This was not what Leclerc said, nor was he giving a definition of liberty. What he said was that we are free to doubt about the truth of propositions which are not evident, a statement that was not at variance with Collins' own declaration that we are necessitated to declare probable propositions probable. There is always some doubt about the truth of probable propositions. He was no more successful with Cheyne. He said Cheyne defined liberty as a power to will evil (knowing it to be evil) as well as good. Again, Cheyne was not giving a definition. All he said was that a man can will something he knows to be evil, in order to show he has freewill. In this case the object of his choice would be the demonstration of his freedom of will and this he would be choosing *sub ratione boni*, even though he would be acting a little perversely. King, said Collins, defined liberty as a power to overcome our reason by the sheer force of our will, so that it could compel it to admit what is false to be true. The ambiguities of King's position have been discussed. If his words are to be taken literally they are open to Collins' strictures, but one wonders if King was thinking not of factual propositions but of value judgements in which our will, influenced by habit and prejudice, can make us think that what is evil and therefore "false" is really good and therefore "true." This section of Collins work is not his most successful.

He returned to King, to a criticism of his theory of indifference. He took King as holding that all the good in objects comes from our choosing them and he himself took up the position that objects have objectively good qualities. He said King's theory would prevent man learning from experience, that it would subject him to many wrong choices and make his senses, appetites and reason to be of no use at all. All this is quite valid criticism, but King's theory, as has been seen, was rather complicated and, it would seem, contradictory. It was certainly not typical of the common theory of liberty of indifference. Collins' attempt to show the imperfection of freewill by an examination of what he described as definitions of freewill can hardly be said to have proved his point.

He went on to attempt to show the perfection of necessity. Now here, paradoxically, the opponents of Collins would agree with him: where they would not agree with him is in the way in which this perfection is explained. He said, and all Christian theologians would agree, that God is necessarily perfect, that he knows all truth and that he is necessarily happy. There would be a difference of opinion with regard to Collins' saying "Is it not a perfection in him to will and to do always what is

best?" "If all things are indifferent to him," continued Collins, "he must will without reason or cause." Leibniz has a rather similar remark. "If they are not indifferent," said Collins, "he must be necessarily determin'd by what is best." Leibniz had God obliged by his goodness to create the best of all compossible worlds.[97] Collins either seems to have God compelled by some cause – even if an ideal cause – outside himself or, by a process of reasoning, involving ideas, within himself. This would imply at least composition in God. But to be absolutely perfect God must be absolutely simple. Collins gives a long quotation from Burnet, from which it might seem that God had necessarily to create. But God, Burnet said, is not obliged to create and all Christian theologians would agree with him. He creates out of his own generosity. He is under no compulsion. But here is a mystery to which Burnet pointed and to which Collins drew the attention of his readers, which human reason cannot solve. God, being totally simple, his act of creation must be one with his Essence and therefore, like his Essence, necessary. Why God created is a mystery precisely because the Essence of God cannot be plumbed by human reason. His act is at once free and necessary. But necessary is here taken in a sense that is not opposed to free. His act, which is one with his Essence, is necessary, in that it cannot not be. It is free in that he is free from all compulsion, even compulsion that comes from his Divine Nature. The question involves theology, but it was Collins who introduced a theological note, in spite of his saying in his Preface that he would keep to purely philosophical arguments. At least it can be regarded as natural theology and Collins avoided the vexed questions that involve grace and freewill.

He went on to deal with the angels and the blessed in heaven. They, he said, quoting Bramhall, are necessitated and their state is more perfect than ours. Hobbes made the same point, and so did Bayle. Bramhall replied that good angels have the freedom to choose between good acts, moreover, "the understandings of the angels are clearer, their power and dominion over their actions is greater, they have no sensitive appetites to distract them, no organs to be disturbed."[98] He missed the main point. Their greater perfection is due not to the nature of their wills but to the fact that the one object that necessitates them – their supreme good, God – is present before them. This state of necessity is more perfect than that of men on earth, but it is the circumstances that make it more perfect. The will, would say the protagonists of liberty of indifference, is still "indifferent" to anything that is not the highest good, but, seeing that highest

---

[97] *Theodicy, Observations*, p. 431.
[98] Hobbes, *English Works*, V, pp. 57, 60, 247, 249; Bayle, *Oeuvres*, III, p. 664; *Réponse*, II, c. 82.

good and the necessary connection of their actions with that good, they cannot reject it or turn away from it in their actions. But the root of indifference remains. It is the circumstances that have changed. Man's freedom of will in this world is not impugned by the argument, and the quality that causes it remains in the next. Men in the next life are not, as Collins said, simply determined by their reasons. They are determined by the *summum bonum* presented to their reason. Collins quoted the passage from Bayle in which the latter said that if man had the quality of indifference "il sera le plus indisciplinable de tous les animaux." He, like Bayle, misunderstood the meaning of indifference. There is a powerful passage in which he argued "is not man more perfect, the more capable he is of conviction?" This is true. He went on, "And will he not be more capable of conviction, if he be necessarily determin'd in his assent by what seems a reason to him, and necessarily determin'd in his several volitions by what seems good to him?" Even if he falls into error, it were better, he said, that he were so necessitated. Now it is irrefutably true that a man would be more perfect if he always saw where the better course lay and always chose that course. But the libertarians would say that this would only be the case if he were able to choose this course freely. If he were under some compulsion which made him consider limited good objects as necessary to him they would say he would be acting against his nature and not as a human being. The angels in heaven, they would say – and again it was Collins who had introduced this theological note – are necessitated precisely because they see the unlimited and perfect good and the connection of limited good objects with that perfect good. In so far as they are necessitated they are acting according to their intellectual natures. Collins did not introduce, as has been said, the question of grace and freewill. That would have raised the problem into another and even more complicated dimension. Collins had again produced a powerful argument for psychic determinism but, in fact, his introduction of the case of the angels can serve to bring out the difference between his position and that of the defenders of liberty of indifference. The angels would produce no argument for his case.

The next proof of determinism is again theological, taken from the Divine foreknowledge of men's free choices. This again is paralleled in Hobbes.[99] God, said Collins, knows what man will choose. Therefore that choice will certainly be made. That would not be questioned. Collins went on to say that God must obtain his knowledge either from his decrees or from the fact that choice is determined in pre-existing causes.

---

[99] Hobbes, *English Works*, IV, p. 270 sq., V, pp. 339, 340.

In either case the choice is not only certain but also necessary, as God's decrees, owing to God's omnipotence, must be necessary. It is a strong argument and, as Collins showed, caused Luther to deny freewill. Scholastic theologians would have said that God knows all things, also, actual and possible, in his Essence, since they are its image and it their ultimate cause, but they would admit he knows them in his decrees. This, however, they would say, does not destroy freewill. God's omnipotence demands that all other beings are created, but he creates them with certain powers, which are also created. It is required that he keeps them in being, and that he uses his creative power to bring into being anything that results from the action of his creatures, using the powers he has given them. But unless the whole world is a machine, acting under dead causal necessity, living creatures must act as living creatures, and free as free, even though they depend for their existence and powers on the creative and conservative power of God. His omnipotence, therefore, is saved, while the creative powers of secondary agents, in this case free beings, is not destroyed. But the mystery of the Divine Nature is involved in the whole question, and Collins was able to quote Christian Divines – Tillotson and Stillingfleet, for example, – who admitted that there remained an element of mystery which they could not resolve.

The last two arguments are based on legal and moral considerations. Again, Collins was using the policy that attack is the best defence. Laws, he said, would be useless if men were not moved by pains and penalties. His morality, taken from Locke, to whom he referred, is hedonistic. "Morality or virtue, consists in such actions as are in their own nature, and upon the whole pleasant: and immorality or vice, consists in such actions as are in their own nature and upon the whole painful." As has been seen, Locke modified this position in other passages in his *Essay*. But, said Collins, if man is indifferent to pain and pleasure he will neither respond to the penalties of the law nor be able to observe morality. Hedonism as the basis of a moral theory can be questioned, but what Collins was certainly doing was again misunderstanding the meaning of "indifference." Pains, pleasures and arguments of reason would certainly affect the man who was endowed with "indifference" in the sense of the defenders of the theory of liberty of indifference.

There followed a number of objections to determinism. The first three were taken directly from Hobbes and dealt with the arguments for freedom taken from law, morality and the use of argument. The objections were first raised by Bramhall in the discussion between him and Hobbes. If man is determined in all his actions, said Bramhall, then any punishment for a breach of the law would be unjust. Not so, said Collins, for the sole

purpose of the law is to remove vicious members from society, and to act as a deterrent. This was Hobbes' point of view. For this reason, said Collins, furious madmen in some parts of the world are the objects of judicial punishment; while men afflicted with plague can be "cut off from society" – the aim in both cases being to remove diseased members as "a canker'd branch is from a tree." Neither of these actions, however, could be regarded as a punishment. Detention in Broadmoor or quarantine are regarded as protection for society, not as punishment. "The Law," wrote Collins, "very justly and rightly regardeth only the will, and no other preceding causes of action." He followed this with a passage that is practically a paraphrase from Hobbes, though without an acknowldegment – Hobbes was not a very respectable author to quote – to the effect that if a man is, by strength of temptation, necessitated to theft, he can justly be hanged, since his theft was voluntary – that is, willed, though not freely willed – and his punishment will therefore be a deterrent to others.[100] Even the children of rebel parents can justly "suffer in their fortunes," he says, "so little is free-agency requisite to make punishments just." Whatever was the case of the law in the eighteenth century, it would not now be regarded as just to punish children for their parents' crimes – though the confiscation of rebel parents' estates could be regarded more as a punishment of the parents than of the children – and there is in law a rule of diminished responsibility. Moreover, deterrence is not the only purpose of a just law. In spite of the unpopularity of the idea of vindicative justice today, it is a necessary element in the just punishment of crime. One does not exact an excessive penalty from a defaulter even though such a penalty would be a more effective deterrent. There must be some proportion between punishment and the degree of culpability of the crime. The motive of deterrence may be useful for fixing the nature of the punishment, but deterrence by itself would not justify punishment. Ignorance of the law may be regarded as no excuse, but to some extent the culpability of this ignorance would be taken into account and at least the law must have been adequately promulgated. Just punishment implies culpability and Collins' theory of justice as it is expressed here does not allow for this last element. Not would it allow for that of reformation.

The second objection turns from the question of the justice of punishment to that of its usefulness. It is useful as a deterrent, Collins said, even when inflicted on those men who act from necessity. This argument is reinforced by the fact that animals, children and madmen may be punished to make them leave off vicious habits. The three cases are very

---

[100] Hobbes, *English Works*, IV, p. 253.

different. Collins' attitude to madmen was not exactly liberal, but he had
a point in saying that animals are regarded as necessitated and punishment can be useful in their case. With regard to praise, blame and argument – which Bramhall had asserted would be of no value when applied
to necessary beings – Collins said that they are "necessary causes to determine certain men's wills to do what we desire of them," but – his old
mistake – neither they nor punishments would be of use if men had liberty
of indifference. He added – and both Hobbes, Bayle and Leibniz gave
the same example – that the man who is – necessarily – virtuous by
nature is regarded as more praiseworthy than he who acquires virtue by
reasoning, which "is a very precarious thing."[101]

Here one has got to ask oneself against what sort of determinism
Collins' list of objections is aimed and what sort of determinism he was
defending. A psychopath, whose actions would be, in some fields at least,
determined, might well not be moved by the sort of arguments, praise,
blame or punishment that would affect more normal men. Collins'
rational but compulsive hedonist, who might be considered to conform
to the pattern of the first edition of Locke's *Essay*, would certainly take
into account rewards, punishments, praise, blame and reasonings. But
Collins in his answer to the objection appeared to be leaning to a Hobbesian and therefore mechanistic view of freewill, and he took a Hobbesian
line in considering such arguments to be of no effect. However, regretfully
perhaps, one would find it difficult to see how one could punish even the
compulsive hedonist justly, though one would have to admit that one
might well do so usefully.

The fifth objection brings the matter to a head. He asks "How a man's
conscience can accuse him if he knows he acts necessarily, and also does
what he thinks best when he commits any sin?" Collins defines conscience as "a man's own opinion of his actions with relation to some
rule." Collins here seems to be departing from hedonism. Locke, in spite
of the subjective element in his definition of conscience as "our own
opinion or judgement of the moral rectitude or pravity of our own actions,"[102] elsewhere in his *Essay*, as has been said, considerably modified
the hedonism that appeared in his definition of good and evil. Collins
said that a man is necessitated, with reluctance, to break that rule. Later
he judges his action to be contrary to it, and "by the absence of the pleasure
of the sin, and by finding himself obnoxious to shame, or by believing

---

[101] *Ibid.*, pp. 255, 256; Leibniz, *Theodicy*, p. 163; Bayle, *Oeuvres*, III, p. 666; *Répon-se*, II, c. 83.

[102] *Essay*, I, 2, no. 8. Cf. Fraser's edition, I, p. 71 for different wordings in different
editions of the *Essay*. This is that of the fourth edition.

himself liable to punishment, he may really accuse himself." Hobbes said repentance is "a glad returning into the right way, after the grief of being out of the way."[103] But if one was necessarily out of the way, how can one blame oneself for this aberration? It is impossible, on Collins' theory, to see how one can really regard oneself as having behaved culpably.

The fourth and sixth objections raise again the query, what sort of determinism was Collins teaching? In the earlier part of the work he had seemed to be close to that of the first edition of Locke's *Essay* and of Leibniz. In all these objections, however, he seems to move closer to Hobbes. What use he asked, by way of objection, will care or physic make to the length of a man's life, if the term of that life be fixed and what harm will want of care do to a man? His reply came to this, that everything in a man's life is determined and is a necessary cause in the chain that will lead to its final end, his death. Care and medicine will be part of such necessary causes. Further he said, "if all events are necessary, it was as impossible for Julius Caesar not to have died in the Senate, as it is impossible for two and two to make six. But who will say that the former was as impossible as the latter is, when we can conceive it possible for Julius Caesar to have died anywhere else as in the Senate and impossible to conceive two and two ever to make six?" His reply is that "if all events are necessary, it was as impossible for Julius Caesar not to have died in the Senate as it is impossible for two and two to make six." Now this goes beyond mechanical determinism. It is sheer logical determinism. Moreover it cannot be a careless remark on Collins' part. It quite obviously has a reference to Bayle's note S, in his article on Chrysippus in which he said, "At present it puzzles the Spinozists, to see, that according to their hypothesis, it was as impossible from all eternity, that Spinoza, for example, should not die at The Hague, as it is impossible, that two and two should be six." He went on, in effect, to say that the death of Spinoza at Leyden would not imply a contradiction. Leibniz approved of this comment, but added that Bayle had continued with a remark "which somewhat spoils his eminently reasonable statement," i.e. "Now what contradiction would there be if Spinoza had died at Leyden? Would nature then have been less perfect, less wise, less powerful?" Leibniz said he was confusing "what is impossible because it implies a contradiction," i.e. a logical contradiction, "with what cannot happen because it is not meet to be chosen." He, a little earlier, said of Hobbes that he had taught an absolute necessity of all things.[104] In his *De Potentia et Actu* Hobbes wrote, "quicumque actus futurus est necessario futurus est, ... Imo vero

[103] Hobbes, *English Works*, IV, p. 257.
[104] *Theodicy*, pp. 234, 235.

non minus necessaria propositio est futurum est futurum quam homo est homo."[105] Collins must certainly have been thinking of the passage in either Bayle or Leibniz or in both. He had the *Theodicy* in his library and, as has been said, he quoted from it or at least from the appended essay on King. It would seem that here he deliberately ranged himself against Bayle and probably against Leibniz. All actions have their surrounding circumstances, he said, which make them necessary with an absolute and indeed a logical necessity. It was a very extreme position. It was the position of Spinoza, a doctrine that particularly alarmed the men of the seventeenth century: the doctrine that held that the logical and ontological orders are the same. Collins had in his library the *Opera Postuma* of Spinoza, which included the *Ethics*. There is no other evidence in his life and writings to connect him with Spinoza. However, in spite of his assertion of logical necessity, his arguments in the *Inquiry* could only be used in an attempt to prove physical necessity. The death of a man at some future date in some place other than that in which he will actually die would not be contrary to the principle of contradiction. But one might well ask whether psychic can not be reconciled with and, in fact, reduced to a rigid causal determinism. If all men's decisions are necessitated by the need to choose whatever in a given set of circumstances seems to be most attractive, is not human choice only another element in an inevitable causal chain, which can only be broken by the Scholastic doctrine of freedom of indifference?

This was the end of Collins' argument. He went on to claim the support of most philosophers, of all those Christians who believe in predestination and, finally, of Samuel Clarke, "whose authority is equal to that of many others put together and makes it needless to cite others after him." The example he chose was taken from Clarke's *Boyle Lectures*, in which Clarke said it is morally impossible for a man, free from all pain of body or worry of mind, or temptation or external pressure, to commit suicide, "not because he wants a natural or physical power to do so but because 'tis ... morally impossible for him to choose to do it."[106] This, Collins said, was precisely his own position. Clarke replied that "by moral necessity, consistent writers never mean any thing more, than to express in a figurative manner the certainty of such an event, as may in reason be fully depended on, though literally and in philosophical strictness of truth there be no necessity at all of the event."[107] He considered, he said, that

[105] Hobbes, *Latin Works*, I, p. 115.
[106] S. Clarke, *A Demonstration of the Being and Attributes of God*, (London, 1728), p. 99.
[107] *Remarks*, pp. 16, 17.

his words, "not because he wants a natural or physical power to do so," sufficiently explained the case.[108] By them, in other words, he meant not bodily power but also power of will. If one reads the whole passage in *The Being and Attributes of God* one can see the distinction drawn quite clearly. Moreover, Clarke caught Collins out in an elementary logical error. Collins said that the "natural power" referred to by Clarke is consistent with and indeed "a consequence of the doctrine of necessity. For, if man is necessarily determined by particular moral causes, and cannot then possibly act contrary to what he does, he must under opposite moral causes, have a power to do the contrary." Clarke's riposte was, "That is to say: a man's having, under the *same* moral causes (which is evidently the sense of Dr. Clarke's words) a physical power to do the contrary to what he does: means only his having, under the *opposite* moral causes, a physical power to do the contrary to what he does."[109] Clarke himself, however, was, on what should have been his own theory, guilty of an inaccuracy. He compared – Collins gives the quotation – the moral necessity affecting the man who cannot commit suicide to that by which the angels cannot do evil. Now the angels cannot do evil because they cannot reject the *summum bonum*, but the man in question is not so necessitated, as he has not the *summum bonum* before him. The cases and the two necessities are not the same. Admittedly, Clarke said the angels have "a perfect knowledge of what is best" and the man can only not commit suicide because "'tis absurd and mischievous." In spite of Collins' error, however, he had picked a strong example, if a very exceptional one. Is a man free in such extreme circumstances? Clarke could have given his position some consistency by saying that even though the man does not see the *summum bonum* he can only choose an object *sub ratione boni*, and in the example given there would be no *ratio boni* whatsoever in the act of suicide, in the circumstances stated, unless the man were to kill himself with the perverse intention of demonstrating the freedom of his will. The conditions as stated come very near to making the object, suicide, no suitable object for the will at all, even though the will retains its power of indifference. The case, therefore, could be fitted into the theory of indifference, which says that man is only necessitated by the highest good and also that he can only will what has some aspect of good about it. What Clarke chose to do instead was to change his ground, to talk, not about moral necessity,[110] but of moral certainty and to say that

---

[108] *Ibid.*, p. 18.
[109] *Ibid.*, p. 18.
[110] Clarke was using the term, not in the sense of moral obligation, but of impulse to act or refrain from acting.

it expresses in a figurative manner the certainty of an event. There is a difference between moral certainty and moral necessity. The first refers to a matter of fact about which one is quite certain, the second to an action which may be free. It would be correct to say that moral certainty is more absolute than the certainty associated with moral necessity. Clarke, however, seems to be using the term in the milder sense of "in reason, extremely probable." In the seventeenth century moral certainty was at times taken to have a much stronger meaning than this. It was taken to apply, for example, to the certainty we have that there is such a place as America.[111] One has perhaps not quite the same degree of certainty, in the circumstances, that one's friend will not commit suicide, even though, as Clarke said, one may "fully depend on it." It is important to get the distinction clear, otherwise one might think that the certainty one has about moral necessity is more absolute than it is. Moral certainty is not the same as logical certainty. Moral necessity is often a "metaphorical expression," meaning no more than extreme probability. In the case in question – that of suicide – it could mean more, on the supposition, and only on the supposition, that there was no *ratio boni* in the act.

Finally the book ended with a definition of man's liberty as "a power to do as he wills, or pleases." This was the definition of Locke in all the editions of his *Essay* and that of Hobbes.[112] Freewill, in fact, is simply freedom from coercion.

CONCLUSION

What is the value of the book? It has faults. It is inclined to be prolix. There are too many quotations – though one reason for this was that Collins wanted to use his opponents' own words as weapons against themselves. Quotations are at times misused or perhaps given a tendentious interpretation. Collins confused judgement and volition and misunderstood the meaning of "indifference" – though here he erred in good company – and his logic could be faulted by Clarke. On the other hand, his book is a very complete defence of determinism. It produces a com-

---

[111] Cf. J. Tillotson, *Works .... Containing fifty four sermons etc.*, (9th ed. London, 1728), pp. vi, 559. In the first passage Tillotson says moral certainty can "be sometimes taken for a high degree of probability," but continues, "it is also frequently us'd for a firm and undoubted assent to a thing upon such grounds as are fit fully to satisfy a prudent man, and in this sense I have always used this term." The example of the existence of America, given on p. 559, is clearly meant to be such a certainty.

[112] *English Works*, IV, p. 273. Locke, *Essay*, II, 21, n. 27. It is from the very next section (n. 28) onwards that occur the considerable alterations in the second edition of the *Essay*. Cf. Fraser's edition, I, p. 329 sq.

bined battery of arguments against freewill, and it would seem that Collins attempted a synthesis of the arguments for psychic, causal and logical determinism. That he was aware that there were differences between the three can be seen from his deliberate insertion of the example of the necessity of the place of the death of Julius Caesar, in his answer to his sixth objection. In his psychic determinism he was most akin to Locke, in the first edition of his *Essay*. His arguments in its favour were a most forceful and effective exposition of the theory. In the metaphysical and theological parts of his work he was more influenced by Hobbes. In his deliberate forays into the ground usually claimed by his opponents, he owed a good deal to Bayle. How far his synthesis was successful is open to question. But it is an interesting question. Hobbes also had what was really logical determinism, combined with a theory of mechanical causality. A contingent thing to him was simply something the cause of which we do not know.[113] And he held that man necessarily chooses that which seems to increase his vitality. Collins' contribution was that he tried to combine Locke and Hobbes, though he did indeed challenge Locke's theory of suspense of appetite, as developed in the second edition of Locke's *Essay*, and that he took this synthesis further by adding to it the idea of logical determinism that he may more probably have derived from Bayle's and Leibniz's remarks on Spinoza than from Spinoza's *Ethics* itself.[114] It was not so much that he tried deliberately to reconcile the three philosophers, as that he attempted to utilise, for his own theory, the ideas of all three. The work, therefore, can be said to have originality in this combination of massed argument and in the attempted synthesis. It was also of historical importance. Its effect on Voltaire, the *philosophes* and Priestley cannot be ignored. Its frequent reprinting up to the end of the eighteenth century is witness to the fact that, unlike the rest of Collins' works, it was not quickly forgotten. A great deal has been written on the problem of determinism since the days of Collins, but the *Philosophical Inquiry* deserves a place in the history of the debate on necessity and freewill.

[113] *English Works*, IV, p. 259.

[114] If this were the case – and the *Ethics* was a book particularly hard to understand and, in fact, little understood in England – he might well not have considered all the implications which his statement involved.

TEXT OF THE *PHILOSOPHICAL INQUIRY*
CONCERNING HUMAN LIBERTY

A

PHILOSOPHICAL

INQUIRY

CONCERNING

HUMAN

LIBERTY.

The second Edition corrected.

LONDON:

Printed for R. ROBINSON, *at the*
Golden Lion *in St.* Paul's Church-
Yard. M DCC XVII.

PREFACE.

TOO much care cannot be taken, to prevent being misunderstood and pre-judg'd, in handling questions of such nice speculation, as those of Liberty and Necessity: and therefore, tho' I might in justice expect to be read before any judgment be pass'd on me, I think it proper to premise the following observations.

I. First, tho' I deny liberty in a certain meaning of that word; yet I contend for liberty, as it sig-nifies,

A 2

PREFACE.

nifies, a power in man, to do as he wills, or pleases; which is the notion of liberty maintain'd by ARISTOTLE, CICERO, Mr. LOCKE, and several other Philosophers, antient and modern. And indeed after a careful examination of the best authors who have treated of liberty, I may affirm, that however opposite they appear in words to one another, and how much soever some of them seem to maintain another notion of liberty; yet at the bottom, there is an almost universal agreement in the notion defended by me, and all that they say, when examin'd, will be found to amount to no more,

9. Se-

PREFACE.

2. Secondly, when I affirm necessity; I contend only for what is call'd moral necessity, meaning thereby, that man, who is an intelligent and sensible being, is determin'd by his reason and his senses; and I deny man to be subject to such necessity, as is in clocks, watches, and such other beings, which for want of sensation and intelligence are subject to an absolute, physical, or mechanical necessity. And here also I have the concurrence of almost all the greatest Asserters of liberty, who either expresly maintain moral necessity, or the thing signified by those words.

3. Thirdly, I have undertaken 'to shew, that the notions I advance, are

iv   PREFACE.

are so far from being inconsistent with, that they are the sole foundations of morality and laws, and of rewards and punishments in society; and that the notions, I explode, are subversive of them. This I judg'd necessary to make out, in treating a subject that has relation to morality: because nothing can be true which subverts those things; and all discourse must be defective, wherein the Reader perceives any disagreement to moral truth; which is as evident as any speculative truth, and much more necessary to be render'd clear to the Reader's mind, than truth in all other sciences.

4. Fourthly, I have intituled my discourse, a Philosophical Inqui-ry

PREFACE.   v

ry &c; because I propose only to prove my point by experience and by reason, omitting all considerations strictly theological. By this method I have reduc'd the matter to a short compass: and hope, I shall give no less satisfaction, than if I had consider'd it also Theologically; for all but Enthusiasts, must think true Theology consistent with reason, and with experience.

5. Fifthly, if any should ask, Of what use such a Discourse is? I might offer to their consideration, first, the usefulness of truth in general; and secondly, the usefulness of the truths I maintain towards establishing laws and morality, rewards and punishments in society: but shall con-

CONTENTS.

a

---

vi. PREFACE.

content myself with observing, that it may be of use to all those who desire to know the truth in the questions I handle, and that think examination the proper means to arrive at that knowledge. As for those, who either make no Inquiries at all, and concern not themselves about any speculations; or who take up with speculations without out any examination; or who read only books to confirm themselves in the speculations they have receiv'd; I allow my book to be of no use to them: but yet think they may allow others to enjoy a taste different from their own.

CON-

A

A
PHILOSOPHICAL
INQUIRY
CONCERNING
HUMAN
LIBERTY.

To Lucius.

"Here send you in wri-
"ting my thoughts
"concerning LIBERTY
"and NECESSITY,
"which you have so often desired of
"me; and in drawing them up,
"have had regard to your penetra-
"tion, by being as short as is con-
"sistent with being understood, and
"to your love of truth, by saying
"nothing but what I think true,
"and also all the truth that I ap-

B "prehend

An Inquiry concerning

2

" prehend relates to the subject, with
" the sincerity belonging to the con-
" versation of friends. If you think
" me either too short in any respect,
" or to have omitted the considera-
" tion of any objection, by its not
" occurring to me, or, that you think
" of importance to be consider'd; be
" pleas'd to acquaint me therewith,
" and I will give you all the satisfa-
" ction I can.

Intro-
duction.

'Tis a common observation, even
among the learned, that there are
certain matters of speculation, about
which it is impossible, from the nature
of the subjects themselves, to speak
clearly and distinctly. Upon which
account, men are very indulgent to,
and pardon the unintelligible dif-
courses of Theologers and Philofophers, which treat of the sublime
points in theology and philosophy.
And there is no question in the whole
compass

Human Liberty,

3

compass of speculation, of which men
have written more obscurely, and of
which it is thought more impoffible to discourse clearly, and concerning which men more expect,
and pardon obscure discourse, than
upon the subjects of *Liberty* and *Necessity*. But this common observation, is
both a common and a learned error.
For whoever employs his thoughts either about God or the Trinity in U-
nity, or any other profound subject,
ought to have some Idea's, to be the
objects of his thoughts, in the same
manner as he has in thinking on the
most common subjects: for where
Idea's fail us in any matter, our
thoughts must also fail us. And it
is plain, whenever we have Idea's, we
are able to communicate them to o-
thers by words † : for words being
B 2   arbi-

† I do not mean unknown simple Idea's. These can at
first only be made known by application of the object to
the faculty: but when they have been once perceiv'd,
and a common name agreed upon to signify them, they
can be communicated by Words.

arbitrary marks of our Idea's, we can never want them to signify our Idea's, as long as we have so many in use among us, and a power to make as many more as we have occasion for. Since then we can think of nothing any farther than we have Idea's, and can signify all the Idea's we have by words to one another; why should we not be able to put one Idea into a Proposition as well as another? Why not to compare Idea's together about one subject as well as another? And why not to range one sort of Propositions into order and method, as well as another? When we use the term GOD, the Idea signify'd thereby, ought to be as distinct and determinate in us, as the Idea of a triangle or a square is, when we discourse of either of them; otherwise, the term GOD is an empty found. What hinders us then from putting the Idea signify'd by the term GOD

into

into a Proposition, any more than the Idea of a triangle or a square? And why cannot we compare that Idea with another Idea, as well as any two other Idea's together; since comparison of Idea's consists in observing wherein Idea's differ, and wherein they agree; to which nothing is requisite in any Idea's, but their being distinct and determinate in our Minds? And since we ought to have a distinct and determinate Idea to the term GOD, whenever we use it, and as distinct and determinate as that of a triangle or a square; since we can put it into a Proposition; since we can compare it with other Idea's on account of its distinctness and determinateness; why should we not be able to range our thoughts about GOD in as clear a method, and with as great perspicuity as about figure and quantity.

I would not hereby be thought

B 3

to

An Inquiry concerning

6

to suppose, that the Idea of GOD is an adequate Idea, and exhausts the subject it refers to, like the Idea of a triangle or a square; or that it is as easy to form in our Minds, as the Idea of a triangle or a square; or that it does not require a great comprehension of Mind to bring together the various Idea's that relate to GOD, and so compare them together; or that there are not several Propositions concerning him that are doubtful, and of which we can arrive at no certainty; or that there are not many Propositions concerning him subject to very great Difficulties or objections. All these I grant: but say, they are no Reasons to justify Obscurity. For, first, an inadequate Idea is no less distinct, as such, than an adequate Idea, and no less true, as far as it goes; and therefore may be discours'd of with equal clearness and truth. Secondly, 'Tho the

Idea

Human Liberty.

7

Idea of GOD be not so easy to form in our Minds as the Idea of a triangle or a square, and it requires a great comprehension of mind to bring together the various Idea's that relate to him, and compare them together; yet these are only reasons, for using a greater application, or for not writing at all. Thirdly, if a writer has in relation to his subject any doubts or objections in his mind, which he cannot resolve to his own satisfaction, he may express those conceptions or thoughts no less clearly, than any other conceptions or thoughts. He should only take care not to exceed the bounds of those conceptions, nor endeavour to make his reader understand what he does not understand himself: for when he exceeds those bounds, his discourse must be dark, and his pains useless. To express what a man conceives is the end of writing; and every reader ought to be satisfy'd,

B 4

An Inquiry concerning

fix'd, when he fees an author fpeak of a fubjeċt according to the light he has about it, fo far as to think him a clear writer.

When therefore any writer fpeaks obfcurely, either about GOD, or any other Idea of his mind, the defeċt is in him. For why did he write before he had a meaning; or before he was able to exprefs to others what he meant? Is it not unpardonable, for a man to cant, who pretends to teach?

Thefe general reflections may be confirm'd by matter of faċt from the writings of the moſt celebrated dogmatical authors.

When fuch great men as GAS-SENDUS, CARTESIUS, CUD-WORTH, LOCKE, BAYLE, Sir ISAAC NEWTON, and Mr. DE FONTENELLE treat of the moſt profound queſtions in metaphyficks, mathematicks, and other parts of philo-

Human Liberty.

philofophy; they by handling them as far as their clear and diſtinct Idea's reach'd, have written with no lefs perfpicuity to their proper readers, than other authors have done about hiſtorical matters, and upon the plaineſt and moſt common fubjeċts.

On the other fide, when authors, who in other refpeċts are equal to the foregoing, treat of any fubjeċts further than they have clear and diſtinct Idea's; they do and cannot but write to as little purpofe, and take as abfurd pains, as the moſt ignorant authors do, who treat of any fubjeċt under a total ignorance, or a confus'd knowledge of it. There are fo many examples of thefe latter occurring to every reader; and there are fuch frequent complaints of mens venturing beyond their ability in feveral queſtions; that I need not name particular Authors, and may fairly avoid the odium of cenfuring any one. But having

An Inquiry concerning

having met with a passage concerning the ingenious Father MALEBRANCHE in the *Letters* of Mr. BAYLE, who was an able Judge, a friend to him and a defender of him in other respects, I hope I may without being liable to exception produce F. MALEBRANCHE as an example. He has in several books treated of and vindicated the opinion of *seeing all things in God*; and yet so acute a person as Mr. BAYLE, after having read them all, declares, that he *less comprehends his notion from his last book than ever* *. Which plainly shows a great defect in F. MALEBRANCHE to write upon a subject he understood not, and therefore could not make others understand.

You see, I bespeak no favour in the

* *J'ai parcouru le nouveau Livre du Pere Malebranche contre Mr. Arnauld; & j'y ai moins compris que jamais sa prétention, que les Idées, par lesquelles nous connoissons les Objets, sont en Dieu, & non dans notre Ame. Il y a là des mal-entendus; ce sont, ce me semble, des equivoques perpetuelles.* Letter of the 16th of October 1705, to Mr. Des Maizeaux.

Human Liberty.

the question before me, and take the whole fault to myself, if I do not write clearly to you on it, and prove what I propose.

And that I may inform you, in what I think clear to myself, I will begin with explaining the sense of the Question.

Man is a *necessary Agent*, if all his actions are so determin'd by the causes preceding each action, that not one past action could possibly not have come to pass, or have been otherwise than it hath been; nor one future action can possibly not come to pass, or be otherwise than it shall be. He is a *free Agent*, if he is able, at any time under the circumstances and causes he then is, to do different things: or, in other words, if he is not unavoidably determin'd in every point of time by the circumstances he is in, and causes he is under, to do that one thing he does, and not possibly to do any other.

The Question stated.

I. This

ment, and thence conclude, they are free, or not mov'd by causes, to do what they do.

They also frequently do actions whereof they repent: and because in the repenting humour, they find no present motive to do those actions, they conclude, that they might not have done them at the time they did them, and that they were *free* from necessity (as they were from outward impediments) in the doing them.

They also find, that they can do as they will, and forbear as they will, without any external impediment to hinder them from doing as they will; let them will either doing or forbearing. They likewise see, that they often change their minds; that they can, and do chuse differently every succes-five moment; and that they frequently deliberate, and thereby are some-times at a near ballance, and in a state of indifference with respect to judging they

15. Argument wherein our Experience is considered.

I. This being a question of fact concerning what we ourselves do; we will, first, consider our own Experience; which if we can know, as sure we may, will certainly determine this matter. And because experience is urg'd with great triumph, by the patrons of *Liberty*, we will begin with a few general reflections concerning the argument of experience : and then we will proceed to our experience itself.

General Reflections on the argument of Experience.

1. The vulgar, who are bred up to believe *Liberty* or *Freedom*, think themselves secure of success, constantly appealing to *Experience* for a proof of their freedom, and being persuaded that they feel themselves free on a thousand occasions. And the source of their mistake, seems to be as follows. They either attend not to, or see not the causes of their actions, especially in matters of little moment,

14   An Inquiry concerning

about some propositions, and willing or chusing with respect to some objects. And experiencing these things, they mistake them for the exercise of *Freedom*, or *Liberty* from *Necessity*. For ask them, whether they think themselves *free*? and they will immediately answer, *Yes*: and say some one or other of these foregoing things, and particularly think they prove themselves *free*, when they affirm, *they can do as they will.*

Nay, celebrated Philosophers and Theologers, both ancient and modern, who have meditated much on this matter, talk after the same manner, giving definitions of *Liberty*, that are consistent with *Fate* or *Necessity*; tho' at the same time they would be thought to exempt some of the actions of man from the power of *Fate*, or thought to assert *Liberty* from *Necessity*. * Ci-CERO defines *Liberty* to be, *a power to do as we will.* And therein several mo- derns

* Opera e. 3068. Ed. Gron.

Human Liberty.   15

derns follow him. One † defines *Li- berty* to be, *a power to act, or not to act, as we will.* Another defines it in more words thus *: *a power to do what we will, and because we will; so that if we did not will it, we should not do it; we should even do the contrary if we willed* ii. And another †, *a power to do or for- bear an actions, according to the deter- mination or thought of the mind, where- by either is preferr'd to the other.* all which definitions, if the Reader will be pleas'd to reflect, he will see 'em to be only definitions of *liberty* or *freedom* from outward impediments of *action,* and not a *Freedom* or *Liberty* from *Necessity*; as I also will shew them to be in the sequel of this Dif- course, wherein I shall contend e- qually with them for such a power as they describe, tho' I affirm, That there is *no Liberty from Necessity.*

ALEXANDER the *Aphrodisean,* most acute Philosopher of the II*d.* Century, Phil. c. 18.

† Placens Eclaircis. sur la Li- berté. p.-

* as we will.

† Lock's Essay of Human Underst. Book II. c. Qu xii. S. &

Fabricii (a Bibl. Gr. Ad. IV. us de Sect.

An Inquiry concerning

Century, and the earlieft commentator now extant upon *Ariftotle*, and efteemed his beft *Defender* and *Interpreter*) defines *Liberty* to be, * *a power to chufe what to do after deliberation and confultation, and to chufe and do what is moft eligible to our reafon; whereas otherwife, we fhould follow our fancy.*

9 Now a choice after deliberation, is a no lefs neceffary choice, than a choice by fancy. For tho' a choice by fancy, or without deliberation, may be one way, and a choice with deliberation may be another way, or different; yet each choice being founded on what is judged beft, the one for one reafon, and the other for another, is equally neceffary; and good or bad reafons, hafty or deliberate thoughts, fancy or deliberation, make no difference.

In the fame manner, † Bifhop BRAMHALL, who has written feveral books for *Liberty*, and pretends to affert

* Defto. preter) defines *Liberty* to be, * a pow-
p. m. 57.

† From Bramhall's Works, p. 735.

Human Liberty.

fert the Liberty taught by ARISTOTLE, defines *Liberty* thus: He fays, *That act which makes a Man's actions to be truly free, is election; which is the deliberate chufing or refufing of this or that means, or the acceptation of one means before another, where divers are*

10 *reprefented by the underftanding.* And that this definition places *Liberty wholly in chufing the feeming beft* p. 697. *means,* and not in chufing the feeming worft means, equally with the beft; will appear from the following paffages. He fays, *actions done in fud-* p. 702. *den and violent paffions, are not free; be-*

11 *caufe there is no deliberation nor election.* — *To fay the will is determined by motives, that is, by reafons or difcourfes, is as much as to fay, that the Agent is* p. 707.

12 *determin'd by himfelf, or is free. Becaufe motives determine not naturally, but morally; which kind of determination is confiftent with true Liberty*—Admitting *that the will follows neceffarily the laft dis-*

C

18 An Inquiry concerning

dictate of the understanding, this is *not destructive of the liberty of the will*; this is only *an hypothetical necessity*. So that 13 Liberty, with him, consists in chusing or refusing necessarily after deliberation; which chusing or refusing is morally and hypothetically determined, or necessary by virtue of the said deliberation.

Lastly, A great *Arminian* Theologer, who has writ a course of *Philosophy*, and enter'd into several controversies on the subject of *Liberty*, makes *Liberty* to consist in * *an indifferency of mind while a thing is under deliberation. For,* says he, *while the mind deliberates, it is free till the moment of action;* because *nothing determines it necessarily to act, or not to act.* Whereas, when the 14 mind has *a thing under deliberation,* that is, when the mind ballances or compares Idea's or motives together, it is then no less *necessarily* determin'd to a state of *Indifferency* by the appearances

* Le Clerc, Bibl. Chois. Tom. xii. p. 103, 104.

Human Liberty. 19

ances of those Idea's and motives, than it is *necessarily determin'd in the very moment of action.* Were a man to be at liberty in this state of *indifference,* he ought to have it in his power to be *not indifferent,* at the same time that he is *indifferent.*

If *experience* therefore proves the *liberty* contended for by the foregoing asserters of *liberty,* it proves men to have *no liberty from necessity.*

2. As the foregoing asserters of liberty, give us definitions of *Liberty,* as grounded on experience, which are confistent with *Necessity;* so some of the greatest Patrons of liberty, do by their concessions in this matter, sufficiently destroy all argument from *Experience.*

ERASMUS in his treatise for *Free-will,* against LUTHER, says, * *That 4 among the difficulties which have exercis'd the Theologers and Philosophers of all ages, there is none greater than the question of* free-

* Oper Tom. 9. p. 1115.

C 2

20   *An Inquiry concerning*

free-will. And Mr. LE CLERC, speak- 15
ing of this Book of ERASMUS, says,
† *that the question of free-will, was too
subtile for* ERASMUS, *who was no Philosopher; which makes him often contradict himself.*

The late Bishop of SARUM ‡, tho' 16
he contends, *Every Man experiences
liberty;* yet owns, *that great difficulties attend the subject on all hands,*
and that therefore *he pretends not to
explain or answer them.*

The famous BERNARD OCHIN, a 17
great *Italian* Wit, has written a most
subtle and ingenious book, intituled,
*Labyrinths concerning Free-will and Pre-
destination,* &c. wherein he shews,
that they who assert, that Man acts
freely, are involv'd in four great difficulties; and that those who assert
that Man acts necessarily, fall into
four other difficulties. So that he
forms eight *Labyrinths,* four against
Liberty, and four against *Necessity.* He
turns

† Bibl.
Choif.
Tom. xii.
p. 51.

‡ Expof.
p. 117.

Pag. 27.

Printed at
Bafil.

Human Liberty.   21

turns himself all manner of ways to
get clear of them; but not being able
to find any solution, he constantly
concludes with a Prayer to GOD to
deliver him from these Abysses. Indeed, in the progress of his work, he
endeavours to furnish means to get
out of this prison: but he concludes,
that the only way, is to say, with
SOCRATES, *Hoc unum scio quod nihil
scio. We ought,* says he, *to rest con-
tented, and conclude, that* GOD *requires
neither the affirmative nor negative of us.*
This is the title of his last chapter,
*Quâ viâ ex omnibus supradictis Labyrin-
this citò exiri possit, quæ doctæ ignorantiæ
viis vocatur.*

A famous Author*, who appeals to 18
common experience, for a proof of *liberty,* confesses, that *the question of liberty is the most obscure and difficult question
in all Philosophy : that the learned
are fuller of contradictions to themselves,
and to one another, on this, than on any
other*

* King
de Orig.
Mali, p. 91.
de liber.
p. 107.

C 3

Pag. 99. *other subject*: And that he writes *a-* Pag. 105. *gainst the common notion of liberty*, and Pag. 117. endeavours to establish *another notion*, which he allows to be *intricate*. 19

But how can all this happen in a plain matter of fact, suppos'd to be experienc'd by every body? What difficulty can there be in stating a plain matter of fact, and describing what every body feels? What need of so much *Philosophy*? And why so many contradictions on the subject? And how can all men experience *Liberty*, when it is allow'd, that the *common notion of* liberty *is false*, or not experienc'd; and *a new notion* of *Liberty*, not thought on before (or thought on but by few) is set up as matter of experience? This could not happen, if matter of fact was clear for liberty.

3. Other Asserters of *Liberty* seem driven into it on account of suppos'd inconveniencies attending the doc-

trine

trine of *Necessity*. The great EPISCOPIUS, in his *Treatise of Free-will*, acknowledges in effect, that the asserters of *Necessity* have seeming experience on their side, and are thereby very numerous; † *They*, as he observes, *alledge one thing of moment, in which they triumph, viz.* "that the "will is determin'd by the under-"standing: *and assert*, that unless "it were so; the will would be a "blind faculty, and might make evil, "as evil, its object; and reject what "is pleasant and agreeable: And "by consequence, that all persua-"sions, promises, reasonings and "threats, would be as useless to a "Man as to a stock or a stone." *This*, he allows to be very *plausible*, and to *have the appearance of probability*; to be *the common sentiment of the schools*; to be *the rock on which the ablest defenders of liberty have split*, *without being able to answer it*; and to

† Opera Vol. I. p. 198, 199, 200.

be

24   *An Inquiry concerning*

be *the reason*, or argument (or rather the matter of experience) *which has made men in all ages, and not a few in this age, fall into the opinion of the fatal necessity of all things.* But *because it makes all our actions necessary, and thereby,* in his opinion, *subverts all religion, laws, rewards and punishments;* he concludes it *to be most certainly false:* and *religion makes him quit this common and plausible opinion.* Thus also 20 many other strenuous Asserters of Liberty, as well as himself, are driven by these supposed difficulties, to deny *manifest experience*. I say, *manifest experience,* for are we not manifestly determin'd by pleasure or pain, and by what seems reasonable or unreasonable to us, to judge or will, or act? Whereas could they see that there are no grounds for laws and morality, rewards and punishments, but by supposing the doctrine of *Necessity*; and that there is no foundation for laws and

Human Liberty.   25

and morality, rewards and punishments, upon the supposition of man's being a free agent, (as shall evidently, and demonstratively appear) they would readily allow experience to be against *Free-will*, and deny *Liberty*, when they should see there was no need to assert it, in order to maintain those necessary things. And as a farther evidence thereof, let any man peruse the discourses written by the ablest authors for liberty, and he will see (as they confess of one another) that they frequently contradict themselves, write obscurely, and know not where to place *Liberty*; at least, he will see that he is able to make nothing of their discourses; no more than * *Letters, p. 521.*
* Mr. LOCKE was of this treatise of 21 EPISCOPIUS, who in all his other writings, shews himself to be a clear, strong, and argumentative writer.

4. There are others, and those contenders for *Liberty*, as well as deniers

56

An Inquiry concerning

denyers of it, who report the persuasions of Men, as to the matter of fact, very differently, and also judge very differently themselves about the fact, from what is vulgarly believed among those who maintain *Free-will.*

An ancient author speaks thus †: *Fate,* says he, *is sufficiently proved from the general receiv'd opinion and persuasion of Men thereof. For, in certain things, when Men all agree, except a few, who dissent from them on account of maintaining some doctrines before taken up, they cannot be mistaken. Wherefore* ANAXAGORAS, *the* Clazomenian, *tho' no contemptible Naturalist, ought not to be judged to deserve any regard, when opposing the common persuasion of all Men he asserts,* " That nothing is done by " Fate; but that it is an empty " name." And according to all authors, recording the opinions of men in this matter, the belief of *Fate,* as to

22

† Alexander de Fato. p. 10.

Human Liberty.

57

to all Events, has continued to be the *most common persuasion,* both of Philosophers and People; as it is at this day *the persuasion* of much the greatest part of mankind, according to the relations of Voyagers. And tho' it has not equally prevail'd among Christians, as it has and does among all other religious parties; yet it is certain, the Fatalists have been, and are very numerous among Christians: and the free-will-Theologers themselves allow, * *That some Christians are as great Fatalists, as any of the antient Philosophers were.*

The acute and penetrating Mr. BAYLE, reports the fact, as very differently understood by those who have thoroughly examin'd and consider'd the various actions of Man, from what is vulgarly suppos'd in this matter. Says he, † *They who examine not to the bottom what passes within them, easily persuade themselves, that they are*

free:

23

* Reeve's Apol. vol. i. p. 150. Sherlock of Prov. 2d. edit.

28   An Inquiry concerning

free : but, they who have considered with care the foundation and circumstances of their actions, doubt of their freedom, and are even perswaded, that their reason and understandings are Slaves that cannot resist the force which carries them along. He says also, in a familiar Letter, *That the best proofs alledg'd for Liberty are, that without it, Man could not sin; and that God would be the author of evil, as well as good thoughts* *.

And the celebrated Mr. LEIBNIZ, that universal genius, on occasion of Archbishop KING's *appeal to experience,* (in behalf of his notion of liberty, viz. *A faculty, which, being indifferent to objects, and over-ruling our passions, appetites, sensations, and reason, chuses arbitrarily among objects; and renders the object chosen agreeable, only because it has chosen it*) denies, that we experience such, or any other *Liberty*; but contends that we rather experience a determination in all our actions. Says he, *We experience*

* Letter of the 13 of December, 1696. to the Abbot Du Bos.

De Orig. mali. c. 5.

24

25

Human Liberty.   29

perience something in us which inclines us to a choice; and if it happens that we cannot give a reason of all our inclinations, a little attention will show us, that the constitution of our bodies, the bodies encompassing us, the present, or preceding state of our minds, and several little matters comprehended under these great causes, may contribute to make us chuse certain objects, without having recourse to a pure indifference, or to I know not what power of the Soul, which does upon objects, what they say colours do upon the Cameleon. In fine, he is so far from thinking that there is the least foundation from Experience, for the said notion of *Liberty*, that he treats it as a *chimera,* and compares it to *the magical power of the Fairies to transform things.*

Lastly, The Journalists of *Paris* are very far from thinking Archbishop KING's notion of liberty to be matter of experience, when they say, *That Dr.* KING, *not satisfy'd with any of the former*

Remarques sur le liv. de l'Orig. du mal. p. 76.

26

Pag. 84.

30 An Inquiry concerning

former notions of Liberty, *proposes a new notion; and carries indifference so far, as to maintain, that pleasure is not the motive but the effect of the choice of the will;* placet res quia eligitur, non eligitur quia placet. *This opinion,* add they, *Journal makes him frequently contradict himself**. 27

So that upon the whole, the affair of of experience, with relation to *liberty*, stands thus. Some give the name Liberty to actions, which when describ-ed, are plainly Actions that are necessary; Others, tho' appealing to vulgar experience, yet inconsistently therewith, contradict the vulgar experience, by owning it to be *an intricate matter*, and treating it after an intricate manner; Others are driven into the defence of *Liberty*, by difficulties imagin'd to flow from the doctrine of *Necessity*, combating what they allow to be matter of seeming experience; Others, and those the most discerning, either think liberty cannot

* Journal des Sçavans of the 16 of March 1705.

Human Liberty. 31

cannot be prov'd by experience, or think Men may see by experience, that they are *necessary Agents*; and the bulk of Mankind have always been persuaded that they are necessary Agents.

Having thus pav'd the way by shewing that liberty is not a plain matter of experience, by arguments drawn from the asserters of liberty themselves, and by consequence subverted the argument from experience for liberty; we will now run over the various actions of Men which can be conceiv'd to concern this subject, and examine, whether we can know from experience, that Man is a free or a necessary Agent. I think those actions may be reduced to these four: 1. Perception of Propositions. 2. Judging of Propositions. 3. Willing. 4. Doing as we will.

1. *Perception of Ideas*. Of this there can be no dispute but it is Ideas.

[margin: Our experience itself consider'd]

[margin: Passage tion of Idea's.]

2

An Inquiry concerning

§ 3

a necessary action of man, since it is not even a voluntary action. The Idea's both of sensation and reflection, offer themselves to us whether we will or no, and we cannot reject 'em. We must be conscious that we think, when we do think; and thereby we necessarily have the Idea's of Reflection. We must also use our senses when awake; and thereby necessarily receive the Idea's of Sensation. And as we necessarily receive Idea's, so each Idea is necessarily what it is in our mind: for it is not possible to make any thing different from itself. This first necessary action, the reader will see, is the foundation and cause of all the other intelligent actions of man, and makes them also necessary. For, as a judicious author and nice observer of the inward acti-

* Locke's ons of Man, says truly, * *Temples*
Posth. *have their sacred images, and we see what*
Works. *influence they have always had over a great*
P. 1, 2. *part*

Human Liberty. 33

part of mankind. But is truth, *the Idea's and Images in mens minds, are the* INVISIBLE POWERS *that constantly govern them, and to these, they universally pay a ready submission.*

2. The second action of man is judging of propositions. All propositions must appear to me either self-evident, or evident from proof, or probable, or improbable, or doubtful, or false. Now these various appearances of propositions to me, being founded on my capacity, and the degree of light propositions stand in to me; I can no more change those appearances in me, than I can change the Idea of red rais'd in me. Nor can I judge contrary to those appearances: for what is judging of propositions, but judging that propositions do appear as they do appear? which I cannot avoid doing, without lying to myself: which is impossible. If any man thinks he can judge a proposition,
D

34 An Inquiry concerning

fition, appearing to him evident, to be not evident; or a probable pro- pofition, to be more or lefs probable than it appears by the proofs to be; he knows not what he fays, as he may fee, if he will define his words. The necefity of being determin'd by appearances, was maintain'd by all the old Philofophers, even by the Academicks or Scepticks. CICERO fays, *You muft take from a Man his fenfes, if you take from him the power of affenting; for it is as necefary the mind fhould yield to what is clear, as that a fcale hanging on a ballance, fhould fink with weight laid on it. For as all living creatures cannot but defire what is agree- able to their natures, fo they cannot but affent to what is clear. Wherefore, if thofe things whereof we difpute are true; it is to no purpofe to fpeak of affent. For he who apprehends or perceives any thing, affents immediately. Again, affent not only precedes the practice of vice; but of virtue,*

Academ. Queft. lib. 2.

Human Liberty. 35

virtue, the fteady performance whereof, and adherence to which, depend on what a man has affented to and approv'd. And it is necefary, that fomething fhould appear to us before we act, and that we fhould af- fent to that appearance. Wherefore he who takes away appearances and affent

29 *from man, deftroys all action in him.* The force of this reafoning mani- feftly extends to all the various judg- ments men make upon the appear- ances of things. And CICERO, as an Academick or Sceptick, muft be fup- pos'd to extend necefity to every kind of judgement or affent of man upon the appearances (or as the Greeks call them φανταςια and himfelf the *Vifa*) of things. SEXTUS EMPIRICUS fays, *they who fay, the Scepticks take away appear- ances, have not convers'd with them, and do not underftand them. For we deftroy not the paffions, to which our fenfes find themfelves expos'd whether we will or no, and which force us to fubmit to appear- ances.*

Pyrrhon. Hypot. l. 2. c. 1c.

D 2

36   *An Inquiry concerning*

ances. For *when it is ask'd us,* whether objects are such as they appear? *we deny not their appearances nor doubt of them, but only question, whether the external objects are like the appearances.* 30

Willing. 3. *Willing,* is the third action of man, which I propose to consider. It is matter of daily experience, that we begin, or forbear, continue, or end several actions barely by a thought or preference of the mind, ordering the doing or not doing, the continuing or ending, such or such actions. Thus before we think or deliberate on any subject, or before we get on horse-back, we do prefer those things to any thing else in competition with them. In like manner, if we forbear these actions, when any of them are offer'd to our thoughts: or if we continue to proceed in any one of these actions once begun: or if at any time we make an end of prosecuting them; we do forbear, or continue,

Human Liberty.   37

continue, or end them on our preference of the forbearance to the doing them, of the continuing them to the ending them, and of the ending to the continuing them. This power of the man thus to order the beginning or forbearance, the continuance or ending of any action, is call'd *the will,* and the actual exercise thereof, *willing.*

There are two questions usually put about this matter: *first,* Whether we are at liberty to will, or not to will? *secondly,* Whether we are at liberty to will one or the other of two or more objects?

1. As to the first, *whether we are at liberty to will, or not to will?* it is manifest, we have not that liberty. For let an action in a man's power be propos'd to him as presently to be done, as for example, *to walk;* the will to walk, or not to walk, exists immediately. And when an action in

D 3

in a man's power is propos'd to him to be done to morrow, *as to walk to morrow*; he is no less oblig'd to have some immediate will. He must either have a will to defer willing about the matter propos'd, or must will immediately in relation to the thing propos'd: and one or the other of those wills must exist immediately, no less than the will to walk, or not to walk in the former case. Wherefore in every proposal of something to be done which is in a man's power to do, he cannot but have some immediate will.

Hence appears the mistake of those who *think* men at *liberty to will, or not to will, because*, say they, *they can suspend willing*, in relation to actions to be done to morrow; wherein they plainly confound themselves with words. For when it is said, man is necessarily determin'd to will; it is not thereby understood, that he is determin'd

Locke of Hum. Und. l. 2. c. 21.

determin'd to will or chuse one out of two objects immediately in every case propos'd to him (or to chuse at all in some cases; as whether he *will* travel into *France* or *Holland*), but that on every proposal he must necessarily have some will. And he is not less determin'd to will, because he does often suspend willing or chusing in certain cases: for *suspending to will*, is itself an *act of willing*; it is willing to defer willing about the matter propos'd. In fine, tho' great stress is laid on the case of *suspending the will*, to prove *liberty*, yet there is no difference between that and the most common cases of willing and chusing upon the manifest excellency of one object before another. For as when a man wills or chuses living in *England* before going out of it (in which will he is manifestly determin'd by the satisfaction he has in living in *England*) he rejects the will to go out

D 4   of

40   *An Inquiry concerning*

of *England*; so a man, who suspends a will about any matter, wills doing nothing in it at present, or rejects for a time willing about it; which circumstances of wholly rejecting, and rejecting for a time, make no variation that affects the question. So that willing or chusing suspension, is like all other choices or wills we have.

2. Secondly, let us now see, *whether we are at liberty to will or chuse one or the other of two or more objects.* Now as to this, we will, first, consider, whether we are at liberty to will one of two or more objects wherein we discern any difference: that is, where one upon the whole seems more excellent than another: or where one upon the whole seems less hurtful than another. And this will not admit of much dispute, if we consider what willing is. Willing or preferring, is the same with respect to good and evil,

Human Liberty.   41

evil, that judging is with respect to truth or falshood. It is judging, that one thing is upon the whole better than another, or not so bad as another. Wherefore as we judge of truth or falshood according to appearances; so we must will or prefer as things seem to us, unless we can lye to ourselves, and think that to be worst, which we think best.

An ingenious author expresses this matter well, when he says, " *the que-* " *stion, whether a man be at liber-* " *ty to will which of the two he* " *pleases, motion or rest; carries the* " *absurdity of it so manifestly in it-* " *self, that one might hereby be suffici-* " *ently convinced, that liberty concerns* " *not the will. For to ask, whether a* " *man be at liberty to will either* " *motion or rest, speaking or silence,* " *which he pleases? is to ask, whe-* " *ther a man can will what he will, or* " *be pleas'd with what he is pleas'd* " *with?*

Locke's Essay of Human Und. l. 2. c. 21. sect. 25.

42 An Inquiry concerning

" with ? A question that needs no an-
" swer."

To suppose a sensible being capable of willing or preferring, (call it
as you please) misery, and refusing
good, is to deny it to be really sensible; for every man, while he has
his senses, aims at pleasure and happiness, and avoids pain and misery;
and this, in willing actions, which
are suppos'd to be attended with
the most terrible consequences. And
therefore the ingenious Mr. NORRIS
† very justly observes, *that all who
commit sin, think it as the instant of commission all things consider'd, a lesser e-
vil; otherwise it is impossible they should
commit it*: and he instances in St. PE-
TER's denial of his Master, who, he
says, *judg'd that part most eligible which
he chose; that is, he judg'd the sin of denying his Master, as that present juncture, to be a less evil, than the danger of not denying him, and so chose it.*
Other-

† Theory of Love p. 199.

32

43 Human Liberty.

*Otherwise, if he had been actually thought
is a greater evil, all that whereby it ex-
ceeded the other, he would have chosen
gratis, and consequently have willed evil
as evil*; which is impossible. And
another acute Philosopher observes,
* *that there are in France many new
converts, who go to mass with great reluctance. They know they mortally offend
God, but as each offence would cost them
(suppose) two pistoles, and having rec-
kon'd the charge, and finding that this
fine paid as often as there are festivals
and sundays, would reduce them and their
families to beg their bread, they conclude
it is better to offend God, than beg.*

In fine, tho' there is hardly anything so absurd, but some ancient
philosopher or other may be cited for
it; yet according to PLATO †, *none
of them were so absurd to say that men
did evil voluntarily*; and he asserts,
that it is contrary to the nature of man, to
follow evil, as evil, and not pursue good;
and

* Bayle Reponse aux Quest. Or. vol.3. p.736.

† Opera Edit. Serran. vol.1. f. 345, 346.

33

34

44   An Inquiry concerning

and that *when a man is compell'd to chuse between two evils, you will never find a man who chuses the greatest, if it is in his power to chuse the less*; and that *this is a truth manifest to all.*

35

And even the greateſt modern advocates for liberty allow, that *whatever the will chuſeth, it chuſeth under the notion of good*; and that *the will is good in general, which is the end of all human actions.*

Bramhall's Works. *p. 656, and 657.*

36

This I take to be ſufficient to ſhew, that man is not at liberty to will one or the other of two or more objects, between which (all things confider'd) he perceives a difference; and to account truly for all the choices of that kind, which can be affign'd.

But, fecondly, ſome of the patrons of liberty contend, that we are free in our choice among things indifferent, or alike, as in chuſing one out of two or more eggs; and that in ſuch

Human Liberty.   45

ſuch cafes the man having no motive from the objects, is not neceffitated to chuſe one rather than the other, becauſe there is no perceivable difference between them, but chuſes one by a mere act of willing without any cauſe but his own free act. To which I anſwer, 1. Firſt, by asking whether this and other inſtances like this are the only inſtances wherein man is free to will or chuſe among objects? If they are the only inſtances wherein man is free to will or chuſe among objects, then we are advanc'd a great way in the queſtion; becauſe there are few (if any) objects of the will that are perfectly alike; and becauſe neceffity is hereby allow'd to take place in all cafes where there is a perceivable difference in things, and confequently in all moral and religious cafes, for the fake whereof ſuch endeavours have been us'd to maintain ſo abſurd and inconfiſtent

46

An Inquiry concerning

inconfiftent a thing as *liberty* or *freedom* from *neceffity*. So that liberty is almoft if not quite, reduc'd to nothing, and deftroy'd as to the grand end in afferting it. If thofe are not the only inftances wherein man is free to will or chufe among objects, but man is free to will in other cafes, thefe other cafes fhould be affign'd, and not fuch cafes as are of no confequence, and which by the great likenefs of the objects to one another, and for other reafons make the caufe of the determination of man's will lefs eafy to be known, and confequently ferve to no other purpofe but to darken the queftion, which may be better determin'd by confidering, *whether man be free to will or no in* more important inftances. 2. Secondly, I anfwer, that whenever a choice is made, there can be no equality of circumftances preceding the choice. 37 For in the cafe of chufing one out of two

47

Human Liberty.

two or more eggs, between which there is no perceiveable difference; there is not nor can there be a true equality of circumftances and caufes preceding the act of chufing one of the faid eggs. It is not enough to render *things* equal to the will, that they are equal or alike in themfelves. All the various modifications of the man, his opinions, prejudices, temper, habit, and circumftances are to be taken in and confider'd as caufes of *election* no lefs than the objects without us among which we chufe; and thefe will ever incline or determine our wills, and make the choice we do make, preferable to us, tho' the external objects of our choice are ever fo much alike to each other. And, for example, in the cafe of chufing one out of two eggs that are alike, there is, firft, in the perfon chufing a will to eat or ufe an egg. There is, fecondly, a will to take but one, or one

48 An Inquiry concerning

one first. Thirdly, consequent to these two wills, follow in the same instant chusing and taking *one*; which *one* is chosen and taken most commonly, according as the parts of our bodies have been form'd long since by our wills or by other causes to an habitual practice, or as those parts are determin'd by some particular circumstances at that time. And we may know by reflection on our actions that several of our choices have been determin'd to one among several objects by these last means, when no cause has arisen from the mere consideration of the objects themselves. For we know by experience, that we either use all the parts of our bodies by habit, or according to some particular cause determining their use at that time. Fourthly, there are in all trains of causes, that precede their effects, and especially effects which nearly resemble each other,

Human Liberty 49

other, certain differences undiscernable on account of their minuteness, and also on account of our not accustoming ourselves to attend to them, which yet in concurrence with other causes as necessarily produce their effect, as the last feather laid on breaks the horse's back, and as a grain necessarily turns the ballance between any weights, tho' the eye cannot discover which is the greatest weight or bulk by so small a difference. And I add, that as we know without such discovery by the eye, that if one scale rises and the other falls there is a greater weight in one scale than the other, and also know that the least additional weight is sufficient to determine the scales; so likewise we may know that the least circumstance in the extensive chain of causes, that precede every effect, is sufficient to produce an effect; and also know, that there must be causes of our choice

E

50 *An Inquiry concerning*

choice (tho' we do not or cannot discern those causes) by knowing, *that every thing that has a beginning must have a cause.* By which last principle we are as necessarily led to conceive a cause of action in man, where we see not the particular cause itself; as we are to conceive that a greater weight determins a scale, tho' our eyes discover no difference between the two weights.

But let us put a case of true equality or Indifference, and what I have asserted will more manifestly appear true. Let two eggs appear perfectly alike to a man; and let him have no will to eat or use eggs. (For so the case ought to be put, to render things perfectly indifferent to him; because, if once a *will* to eat eggs be suppos'd, that *will* must necessarily introduce a train of causes which will ever destroy an equality of circumstances in relation to the things which

Human Liberty. 51

which are the objects of our choice. There will soon follow a second will to eat one first. And these *two wills* must put the man upon action and the usage of the parts of his body to obtain his end; which parts are determin'd in their motions either by some habitual practice or by some particular circumstance at that time, and cause the man to chuse and take one of them first rather than the other.) The case of equality being thus rightly stated, I say, it is manifest no choice would or could be made; and the Man is visibly prevented in the beginning from making a choice. For every man experiences, that before he can make a choice among eggs, he must have a will to eat or use an egg; otherwise he must let them alone. And he also experiences in relation to all things which are the objects of his choice, that he must have a precedent

will

E 2

52

An Inquiry concerning

will to chufe; otherwife he will make no choice. No man marries one wo-man preferable to another, or travels into *France* rather than into another country, or writes a book on one fubject rather than another, but he muft firft have a precedent will to marry, travel, and write.

It is therefore contrary to experi-ence, to fuppofe any choice can be made under an equality of circum-ftances. And by confequence it is matter of experience, that man is ever determin'd in his willing or acts of volition and choice.

Doing as we will

4. Fourthly, I fhall now confider the actions of men confequent to *will-ing*, and fee whether he be *free* in any of thofe actions. And here alfo we ex-perience perfect neceffity. If we will thinking or deliberating on a fubject, or will reading, or walking, or riding, we find we muft do thofe actions, unlefs fome external impediment, as

an

Human Liberty.

53

an apoplexy or fome fuch intervening caufe, hinders us; and then we are as much neceffitated to let an action alone, as we were to act according to our will, had no fuch external im-pediment to action happen'd. If al-fo we change our wills after we have begun any of thefe actions, we find we neceffarily leave off thefe actions, and follow the *new will* or choice. And this was ARISTOTLE's fenfe of fuch actions of man. *As fays he, in arguing we neceffarily affent to the inference or conclufion draws from premifes, fo if that arguing relate to practife, we neceffarily act upon fuch in-ference or conclufion. As for example, when we argue thus, whatever is fweet is to be tafted, this is fweet, he who infers, therefore this ought to be tafted, neceffarily taftes that fweet thing if there be no obftacle to hinder him.*

For a conclufion of this argument from experience, let us compare the actions

Ethica. l. 7. c. 5. ap. Oper. Edit. Par. Vol. II. p. 88. 67.

38

F 3

54 *An Inquiry concerning*

actions of inferior intelligent and sensible agents, and those of men together. It is allow'd that beasts are necessary agents, and yet there is no perceivable difference between their actions and the actions of men, from whence they should be deem'd *necessary* and men *free* agents. *Sheep*, for example, are suppos'd to be *necessary agents*, when they stand still, lie down, go slow or fast, turn to the right or left, skip, as they are differently affected in their minds; when they are doubtful or deliberate which way to take; when they eat and drink out of hunger and thirst; when they eat or drink more or less according to their humour, or as they like the water or the pasture; when they chuse the sweetest and best pasture; when they chuse among pastures that are indifferent or alike; when they copulate; when they are fickle or stedfast in their amours; when

Human Liberty. 55

when they take more or less care of their young; when they act in virtue of vain fears; when they apprehend danger and fly from it, and sometimes defend themselves; when they quarrel among themselves about love or other matters, and terminate those quarrels by fighting; when they follow those leaders among themselves that presume to go first; and when they are either obedient to the shepherd and his dog, or refractory. And why should man be deem'd *free* in the performance of the same or the like actions? He has indeed more knowledge than sheep. He takes in more things, as matter of pleasure, than they do; being sometimes mov'd with notions of honour and virtue, as well as with those pleasures he has in common with them. He is also more mov'd by absent things, and things future, than they are. He is also subject to

E   more

56 *An Inquiry concerning*

more vain fears, more mistakes and wrong actions, and infinitely more absurdities in notions. He has also more power and strength, as well as more art and cunning, and is capable of doing more good and more mischief to his fellow-men than they are to one another. But these larger powers and larger weaknesses, which are of the same kind with the powers and weaknesses of sheep, cannot contain liberty in them, and plainly make no perceivable difference between them and men, as to the general causes of action, in finite intelligent and sensible beings; no more than the different degrees of these powers and weaknesses, among the various kinds of beasts, birds, fishes, and reptiles do among them. Wherefore I need not run thro' the actions of *foxes* or any of the more subtile animals, nor the actions of *children*, which are allow'd by the Advocates

Human Liberty. 57

Advocates of *liberty* to be all necessary. I shall only ask these questions concerning the last. To what 40 age do children continue necessary agents, and when do they become *free*? what different experience have they when they are suppos'd to be free agents, from what they had while necessary agents? And what different actions do they do, from whence it appears, that they are *necessary* agents to a certain age, and *free* agents afterwards?

Dodwell's Works, p. 656, 662.

II. A second reason to prove man a necessary agent is, because all his actions have a beginning. For whatever has a beginning must have a 41 ry cause.

1st Argument taken from the impossibility of Liberty.

If any thing can have a beginning which has no cause, then nothing can produce something. And if nothing can produce something, then the

58 *An Inquiry concerning*

the world might have had a beginning without a cause: which is not only an absurdity commonly charg'd on Atheists, but is a real absurdity in itself

Besides, if a cause be not a necessary cause, it is no cause at all. For if causes are not necessary causes; then causes are not suited to, or are indifferent to effects; and the *Epicurean System* of chance is rendred possible; and this orderly world might have been produc'd by a disorderly or fortuitous concourse of atoms; or, which is all one, by no cause at all. For in arguing against the Epicurean system of chance, do we not say, (and that justly) that it is impossible for chance ever to have produc'd an orderly system of things, as not being a cause suited to the effect; and that an orderly system of things, which had a beginning, must have had an intelligent Agent for its cause, as

Human Liberty. 59

as being the only proper cause to that effect? All which implies, that causes are suited or have relation to some particular effects, and not to others. And if they be suited to some particular effect and not to others, they can be no causes at all to those others. And therefore a cause not suited to the effect, and no cause, are the same thing. And if a cause not suited to the effect, is no cause; then a cause suited to the effect is a necessary cause: for if it does not produce the effect, it is not suited to it, or is no cause at all of it.

Liberty therefore, or a power to act or not to act, to do this or another thing under the same causes, is an *impossibility* and *unphysical*.

And as *liberty* stands, and can only be grounded on the absurd principles of *Epicurean Atheism*; so the *Epicurean Atheists*, who were the most popular and most numerous sect of the

60   An Inquiry concerning

the *Atheists* of antiquity, were the
great * asserters of *Liberty*; as on the
other side, the † *Stoicks*, who were the
most popular and most numerous sect
among the religionaries of antiquity,
were the great asserters of fate and necessity. The case was also the same
among the *Jews*, as among the Hea-
then: the *Jews*, I say, who besides the
light of nature had many books of Revelation (some whereof are now lost);
and who had intimate and personal
conversation with God himself. They
were principally divided into three
sects, the *Sadducees*, the *Pharisees*, and
the *Essenes*. The *Sadducees*, who were
esteem'd an irreligious and atheistical
sect, maintain'd *the liberty of man.*
But the *Pharisees*, who were a religious sect, *ascrib'd all things to fate or
to God's appointment, and it was the first
article of their creed, that fate and God do
all*; and consequently they could not
assert *a true liberty*, when they asserted

Margin notes: * Lucre-tius l. 2. v. 250, &c. Euf. Prep. Ev. l. 6. c. 7. — † Cicero de Nat. Deor. l. 1. — 44 — Josephus the Antiq. l. 18. c. 2. — De bello Jud. l. 2. c. 7.

a

Human Liberty.   61

a liberty together with this *fatality and
necessity of all things.* And the *Essenes,* who were the *most religious sect* a-
mong the *Jews,* and fell not under the
censure of our Saviour for their hypo-
crisy as the *Pharisees* did, were asserters of *absolute fate and necessity.* St.
PAUL, who was a *Pharisee and the son
of a Pharisee,* is suppos'd by the learned
DODWELL, *to have received his doctrine
of fate from the masters of that sect, as
they received it from the Stoicks.* And
he observes further, that *the Stoick
Philosophy is necessary for the explication
of Christian Theology*; that *there are examples in the holy scriptures of the holy
Ghost, speaking according to the opinions
of the Stoicks*; and that in particular,
the Apostle St. PAUL *is what he has dis-
puted concerning Predestination and Re-
probation, is to be expounded according to
the Stoicks opinion concerning fate.* So
that *liberty* is both the real foundation of popular Atheism, and has been
the

Margin notes: 45 — Acts 23 6. — Proleg. ad Steam. deObstin. sect. 40, & 41. — 46

An Inquiry concerning

the profess'd principle of the Atheists themselves; as on the other side, *fate* or the *necessity of events*, has been esteem'd a religious opinion and been the profess'd principle of the religious, both among Heathens and Jews, and also of that great Convert to Christianity and great converter of others, St. PAUL.

III. Thirdly, *Liberty* is contended for by the patrons thereof as a great *perfection*. In order therefore to dif-prove all pretences for it, I will now show, that according to all the va-rious descriptions given of it by Theologers and Philosophers, it would often be an *imperfection*, but never a *perfection*, as I have in the last article show'd it to be *impossible* and *athe-istical*.

1. If *liberty* be defin'd, *a power to pass different judgments at the same in-stant of time upon the same individual propo-*

3ᵈ Argu-ment ta-ken from the Im-perfecti-on of Li-berty.

Le Clerc. Bibl. Choif. tom. xii. p. 88, 89.

Human Liberty.

propositions that are not evident (we be-ing, as it is own'd, *necessarily deter-min'd to pass but one judgment on evi-dent propositions*) it will follow, that men will be so far irrational, and by consequence *imperfect* agents, as they have that *freedom of judgment.* For, since they would be irrational agents, if they were capable of judging evi-dent propositions not to be evident; they must be also deem'd irrational agents, if they are capable of judging the selfsame probable or improbable propositions not to be probable or im-probable. The appearances of all pro-positions to us, whether evident, pro-bable, or improbable, are the sole rational grounds of our judgments in relation to them: and the ap-pearances of probable or improbable propositions, are no less necessary in us from the respective reasons by which they appear probable or im-probable, than are the appearances of

of

An Inquiry concerning

of evident propositions from the reasons by which they appear evident. Wherefore if it be rational and a perfection to be determin'd by an evident appearance, it is no less so to be determin'd by a probable or improbable appearance; and consequently an imperfection not to be so determin'd.

It is not only an absurdity, and by consequence an imperfection, not to be equally and necessarily determin'd in our respective judgments, by probable and improbable, as well as by evident appearances, which I have just now proved; but even not to be necessarily determin'd by probable appearances, would be a *greater imperfection*, than not to be necessarily determin'd by evident appearances: because almost all our actions are founded on the probable appearances of things, and few on the evident appearance of things. And there-

Human Liberty.

therefore, if we could judge, that what appears probable, is not probable but improbable or false; we should be without the best rule of action and assent, we can have.

2. Were *liberty* defin'd, *a power to overcome our reason by the force of choice*, as a celebrated Author may be suppos'd to mean, when he says, *the will seems to have so great a power over the understanding, that the understanding being over-rul'd by the election of the will, not only takes what is good to be evil, but is also compelled to admit what is false to be true*; man would, with the exercise of such a power, be the most irrational and inconsistent being, and by consequence, the most *imperfect* understanding being, which can be conceiv'd. For what can be more irrational and inconsistent, than to be able to refuse our assent to what is evidently true to us, and to assent to what we

F

see

* King de orig. maii. p. 131.

66 An Inquiry concerning

fee to be evidently falfe, and thereby inwardly give the lye to the underftanding?

Cheyne's Phil. Prin. c. 3. C. 13.

will evil (knowing it to be evil) as well as good

3. Were *liberty* defin'd, *a power to will evil* (knowing it to be evil) *as well as good*; that would be an Imperfection in man confider'd as a fenfible being, if it be an imperfection in fuch a being to be miferable. For *willing evil*, is chufing to be miferable, and bringing knowingly deftruction on ourfelves. Men are already fufficiently unhappy, by their feveral judgments, and by their feveral volitions; founded on the wrong ufe of their faculties, and on the miftaken appearances of things. But what miferable beings would they be, if inftead of chufing evil under the appearance of good (which is the only cafe wherein men now chufe evil) they were indifferent to good and evil, and had the power to chufe *evil as evil*, and did actually chufe *evil*

in-

Human Liberty. 87

as evil in virtue of that power? They would in fuch a ftate or with fuch a liberty be like Infants that cannot walk, left to go alone, with liberty to fall: Or like Children, with knives in their hands: Or laftly like young rope-dancers, left to themfelves, on their firft effays upon the rope, without any one to catch them if they fall. And this miferable ftate following from the fuppofition of *liberty*, is fo vifible to fome of the greateft advocates thereof, that they acknowledge, that *created beings, when in a ftate of happiness cafe to have liberty* (that is, ceafe to have liberty to chufe evil) *being inviolably attach'd to their duty by the actual enjoyment of their felicity.*

Bibl. Choife. Tom. xii. p. 95.

Bramhall's Works p. 655.

4. Were *liberty* defin'd, as it is by fome, *a power to will or chufe at the fame time any one out of two or more indifferent things*; that would be no perfection. For thofe things call'd here

in-

68 *An Inquiry concerning*

indifferent or alike, may be considered, either as really different from
each other, and that only seem indifferent or alike to us thro' our
want of discernment; or as exactly
like each other. Now the more *liberty* we have in the first kind, that
is, the more instances there are of
things which seem alike to us and
are not alike; the more mistakes and
wrong choices we must run into.
For if we had just notions, we
should know those things were not
indifferent or alike. This *liberty*
therefore would be founded on a
direct imperfection of our faculties.
And as to *a power of chusing* differently at the same time among *things,
really indifferent*; what benefit, what
perfection would such a power of
chusing be, when the things that
are the sole objects of our *free choice*
are all alike?

5. Lastly, a celebrated Author
seems

Human Liberty. 69

seems to understand by liberty,
a faculty, which, being indifferent to ob- King de
jects, and over-ruling our passions, appe- orig mali.
tites, sensations, and reason, chuses arbi- c. s.
*trarily among objects; and renders the
object chosen agreeable, only because it
has chosen it.*

My design here is to consider this
definition, with the same view, that
I have consider'd the several foregoing definitions, *viz.* to show, that
liberty, inconsistent with necessity, however describ'd or defin'd, is an *imperfection.* Referring therefore my reader for a confutation of this *new* notion of *liberty* to the other parts of my
book, wherein I have already prov'd,
that the existence of such an *arbitrary faculty* is contrary to experience,
and impossible; that our *passions, appetites, sensations,* and *reason,* determine us in our several choices; and
that, we chuse objects because they
please us, and not, as the author
F 3   pretends,

70   *An Inquiry concerning*

Pag. 113 pretends, that *objects please us, only because we chuse them :* I proceed to 52 shew the *imperfections* of this last kind of *liberty*.

1. First, the pleasure or happiness accruing from the *liberty* here assert-ed is less than accrues from the *hy-pothesis of necessity*.

All the pleasure and happiness said to attend this pretended *liberty* consists * wholly in † *creating plea-*53*sure and happiness by chusing ob-jects.*

* Page 127, 108.
† P. 107.

Now man, consider'd as an intelli-gent necessary agent, would no less *create* this pleasure and happiness to himself by *chusing objects*; than a be-ing indu'd with the said *faculty:* if it be true in fact, that *things please us, because we chuse them.*

But man, as an intelligent neces-sary agent, has these further plea-sures and advantages. He, by not being indifferent to objects, is mov'd by

Human Liberty.   71

by the goodness and agreeableness of them, as they appear to him, and as he knows them by reflection and experience. It is not in his power to be indifferent to what causes plea-sure or pain. He cannot resist the pleasure arising from the use of his passions, appetites, senses, and rea-son: and if he suspends his choice of an object, that is presented to him, by any of these powers as a-greeable; it is, because he doubts or examines, whether upon the whole the object would make him happy; and because he would gra-tify all these powers in the best man-ner he is able, or at least such of these powers as he conceive tend most to his happiness. If he makes a choice which proves disagreeable, he goes thereby an experience, which may qualify him to chuse the next time with more satisfaction to him-self. And thus wrong choices may turn

F 4

72 An Inquiry concerning

turn to his advantage for the future. So that, at all times and under all circumstances, he is pursuing and enjoying the greatest happiness, which his condition will allow.

It may not be improper to observe, that some of the pleasures he receives from objects, are so far from being the effect of *choice*, that they are not the effect of the least premeditation or any act of his own, as in finding a treasure on the road, or in receiving a legacy from a person unknown to him.   54

2. Secondly, this *arbitrary faculty* would subject a man to more *wrong choices*, than if he was determin'd in his choice.   55

A man, determin'd in his choice by the appearing nature of things, and the usage of his intellectual powers, never makes a wrong choice, but by mistaking the true relation of things to him. But a being, indifferent

rent,

* P. 147, to 150.

Human Liberty. 73

56 rent to † *all objects*, and sway'd by no motives in his choice of objects, chuses at a venture; and only makes

57 a right choice, when ‖ *it happens* (as the author justly expresses his notion) that he chuses *an object*, which he can by his *creating* power render so agreeable, as that it may be call'd a *rightly chosen object*. Nor can this faculty be improv'd by any experience: but must ever continue to chuse at a venture, or as it *it happens*. For if this *faculty*, improves by experience, and will have regard to the agreableness or disagreeableness of objects in themselves; it is no longer the *faculty* contended for, but a *faculty* mov'd and affected by the nature of things.

So that man, with a *faculty* of choice indifferent to all objects, must make more *wrong choices*, than man consider'd as a necessary being; in the same proportion, as *acting as it happens,*

† P. 106, 111.

‖ P. 106, 107, 113, 139, 141, 147.

74

An Inquiry concerning

happens, is a worse direction to chuse right, than the use of our senses, experience, and reason.

3. Thirdly, the existence of such an *arbitrary faculty*, to chuse without regard to the qualities of objects, would destroy the use of our senses, appetites, passions, and reason; which have been given us to direct us in our inquiries after truth, in our pursuit after happiness, and to preserve our beings. For, if we had *a faculty*, which chose without regard to the notices and advertisements of these powers, and by its choice over-ruled them; we should then be indu'd with *a faculty* to defeat the end and uses of these powers.

The Perfection of *necessity.* But the *imperfection* of liberty inconsistent with necessity, will yet more appear by considering the great *perfection* of being necessarily determin'd.

Can

75

Human Liberty.

Can any thing be perfect, that is not necessarily perfect? For whatever is not necessarily perfect may be imperfect, and is by consequence imperfect.

Is it not a perfection in God necessarily to know all truth?

Is it not a perfection in him to be necessarily happy?

Is it not also a perfection in him to will and do always what is best? For if all things are *indifferent* to him, as some of the advocates of li-

King de orig. mali. p. 177.

58 berty assert, and become *good* only by his *willing* them; he cannot have any motive from his own Idea's, or from the nature of things, to *will* one thing rather than another; and consequently he must *will* without

59 any reason or cause: which cannot be conceiv'd possible of any being; and is contrary to this self-evident truth, that *whatever has a beginning*

60 *must have a cause.* But if things are
not

76 An Inquiry concerning

not indifferent to him, he must be necessarily determin'd by what is best. Besides, as he is a wise being, he must have some end and design: and as he is a good being, things cannot be *indifferent* to him, when the happiness of intelligent and sensible beings, depend on the will he has, in the formation of things. With what consistency therefore can those advocates of liberty assert GOD to be *a holy and good being,* who maintain 61

Pag. 117. that *all things are indifferent* to him before he wills any thing; and that he may will, and do *all things,* which they themselves esteem wicked and unjust?

I cannot give a better confirmation of this argument from the consideration of the Attributes of GOD, than by the judgment of the late Bishop of SARUM; which has the more weight, as proceeding from a great asserter of liberty, who by the force 62

Human Liberty. 77

force of truth is driven to say what he does. He grants, that *infinite per-* Expos. *fection excludes successive thoughts is* p. 26, 27. God; *and therefore that the Essence of God is one perfect thought, in which be views and wills all things.* And though *his transient acts such as creation, providence, and miracles, are done in a succession of time; yet his immanent acts, his knowledge and decrees, are one with his essence.* And as he grants this to be a true notion of God, so be allows that *a vast difficulty arises from it against the* liberty of GOD. *For,* says he, *the immanent acts of God being suppos'd free, it is not easy to imagine how they should be one with the divine essence; to which, necessary existence does most certainly belong. And if the immanent acts of God are necessary, then the transient must be so likewise, as being the certain effects of his immanent acts: and a chain of necessary fate must run through the whole order of things:* and

An Inquiry concerning

and God himself then is no free being, but acts by a necessity of nature. And this necessity, to which God is thus subject, is, adds he, no absurdity to some. God is, according to them, necessarily just, true, and good, by an intrinsick necessity that arises from his own infinite perfection. And from hence they have thought, that since God acts by infinite wisdom and goodness, things could not have been otherwise than they are: for what is infinitely wise or good cannot be alter'd, or made either better or worse. And he concludes, that he must leave this difficulty without pretending to explain it, or answer the objections that arise against all the several ways by which Divines have endeavour'd to resolve it.

Bramhal's Works. p. 656, 695.

63

Again, are not Angels and other heavenly beings esteem'd more perfect than men; because, having a clear insight into the nature of things, they are necessarily determin'd

min'd to judge right in relation to truth and falshood, and to chuse right in relation to good and evil, pleasure and pain; and also to act right in pursuance of their judgment and choice? And therefore would not man be more perfect than he is, if, by having a clear insight into the nature of things, he was necessarily determin'd to assent to truth only, to chuse only such objects as would make him happy, and to act accordingly?

64

Further, is not man more perfect, the more capable he is of conviction? And will he not be more capable of conviction, if he be necessarily determin'd in his assent by what seems a reason to him, and necessarily determin'd in his several volitions by what seems good to him; than if he was indifferent to propositions notwithstanding any reason for them, or was indifferent to any objects notwithstanding

withstanding they seem'd good to him? For otherwise, he could be convinc'd upon no principles, and would be the most undisciplinable and untractable of all Animals. All advice and all reasonings would be of no use to him. You might offer arguments to him, and lay before him pleasure and pain; and he might stand unmov'd like a rock. He might reject 65 what appears true to him, assent to what seems absurd to him, avoid what he sees to be good, and chuse what he sees to be evil. Indifference therefore to receive truth, that is, *liberty* to deny it when we see it; and Indifference to pleasure and pain, that is, *liberty* to refuse the first and chuse the last, are direct obstacles to knowledge and happiness. On the contrary, to be necessarily determin'd by what seems reasonable, and by what seems good, has a direct tendency to promote truth and happiness, and

is

is the proper perfection of an understanding and sensible being. And indeed it seems strange that men should allow that God and Angels act more perfectly because they are determin'd by reason; and also allow, that clocks, watches, mills, and other artificial unintelligent beings are the better, the more they are determin'd to go right by weight and measure; and yet that they should deem it a perfection in man not to be determin'd by his reason, but to have liberty to go against it. Would it not be as reasonable to say, it would be a perfection in a clock not to be necessarily determin'd to go right, but to have its motions depend upon chance?

Again, tho' man does thro' weakness and imperfection fall into several mistakes both in judging and willing in relation to what is true and good; yet he is still less ignorant and

G

cessarily determin'd by the greatest evidence to assent to truth, nor by the strongest inclination for happiness to chuse pleasure and avoid pain; to all which it is a perfection to be necessarily determin'd. For if any action whatsoever can be done without a cause; then effects and causes have no necessary relation, and by consequence we should not be necessarily determin'd in any case at all.

IV. A fourth argument to prove man a necessary agent, shall be taken from the consideration of the divine prescience. The divine Prescience supposes, that all things future will certainly exist in such time, such order, and with such circumstances; and not otherwise. For if any things future were contingent, or uncertain, or depended on the liberty of man, that is, might or might not happen; their certain existence

Fourth argument taken from the divine prescience.

66

G 2

---

and less unhappy by being necessarily determin'd in judging by what seems reasonable, and in willing by what seems best, than if he was capable of judging contrary to his reason and willing against his senses. For, were it not so, what *seems false*, would be as just *a rule of truth*, as what *seems true*; and what *seems evil*, as just *a rule of good*, as what *seems good.* Which are absurdities too great for any to affirm; especially if we consider, that there is a perfectly wise and good Being, who has given men senses and reason to conduct them.

Lastly, it is a perfection to be necessarily determin'd in our choices, even in the most indifferent things: because, if in such cases there was not a cause of choice, but a choice could be made without a cause; then all choices might be made without a cause, and we should not be necessarily

84 An Inquiry concerning

iftence could not be the object of the divine prefcience: it being a contradiction to know that to be certain, which is not certain: and God himfelf could only guefs at the exiftence of fuch things. And if the divine prefcience fuppofes the *certain* exiftence of all things future, it fuppofes alfo the *neceffary* exiftence of all things future; becaufe GOD can fore-know their certain exiftence only, either as that exiftence is the effect of his decree, or as it depends on its own caufes. If he fore-knows that exiftence, as it is the effect of his decree; his decree makes that exiftence neceffary: for it implies a contradiction for an all-powerful being to decree any thing which fhall not neceffarily come to pafs. If he fore-knows that exiftence, as it depends on its own caufes; that exiftence is no lefs neceffary: for it no lefs implies a contradiction, that caufes

fhould

Human Liberty. 85

fhould not produce their effects (caufes and effects having a neceffary relation to and dependence on each other) than that an event fhould not come to pafs, which is decreed by God.

CICERO has fome paffages to the purpofe of this argument. Says he, *De Di. vin. c. 2.*

Qui poteft provideri quidquam futurum effe quod neque caufam habet ullam, neque notam, cur futurum fit? — Quid eft igitur, quod cafu fieri aut forte fortunâ, putemus? —— Nihil eft enim tam contrarium rationi & conftantiæ quam fortuna; ut mihi ne in Deum cadere videatur, ut fciat, quid cafu & fortuito futurum fit. Si enim fcit, certè illud eveniet. Sin certè eveniet, nulla eft fortuna. Eft autem fortuna. Rerum igitur fortuitarum nulla eft præfentio.

Alfo that illuftrious Reformer LuTHER fays, in his treatife againf *Cap. 145.* free-will: *Conceffâ Dei præfcientiâ & omnipotentiâ, fequitur naturaliter irrefragabili*

G 3

86 An Inquiry concerning

fragabili consequentiâ, nos per nos ipsos non esse factos, nec vivere, nec agere quicquam, sed per illius omnipotentiam. *Cum autem tales mos ille ante præscierit futuros, talesque nunc faciat, moveat, & gubernet; quid potest fingi quæso, quod in nobis liberum sit, aliter & aliter fieri, quàm ille præscierit, aut nunc agat? Pugnat itaque ex diametro præscientia & omnipotentia Dei cum nostro libero arbitrio. Aut enim Deus falletur præsciendo, errabit & agendo (quod est impossibile) aut nos agemus & agemur secundum ipsius præscientiam & actionem.* And our learned Dr. SOUTH 68 says, *the fore-knowledge of any event does certainly and necessarily infer, that there must be such an event; for as much as the certainty of knowledge depends upon the certainty of the thing known. And in this sense it is, that God's decree and promise give a necessary existence to the thing decreed or promised, that is to say, they infer it by infallible consequence; so that*

Sermons Vol. III. p. 488.

Human Liberty. 87

that it was as impossible for Christ not to rise from the dead, as it was for God absolutely to decree and promise a thing, and yet the thing not come to pass.

I could also bring in the greatest 69 Divines and * Philosophers who are asserters of liberty, as confirming this argument; for * they acknowledge, that they are unable to reconcile the *divine prescience* and the *liberty* of man together: which is all I intended to prove by this argument, taken from the consideration of the *divine Prescience.*

'See among o-thers Cartesii Prin. pars I. Art. 41. Locke's Letters, p. 27. Tillot-son's mons. Vol. VI. p. 157. Stilling-fleet of Christ's satisfacti-on, p. 355.

V. A fifth argument to prove man a necessary agent, is as follows: If man was not a necessary agent, determin'd by pleasure and pain, there would be no foundation for rewards and punishments, which are the † *essential supports of society.*

Fifth argument taken from the nature of rewards and punishments

† Solon rempublicam contineri dicebat duabus rebus, præmio & pœnâ. Cicero Epist. 15. ad Brutum.

For

G 4

88　An Inquiry concerning

For if men were not neceffarily determin'd by pleafure and pain, or if pleafure and pain were no caufes to determine mens wills; of what ufe would be the profpect of rewards to frame a man's will to the obfervation of the law, or punifhments to hinder his tranfgreffion thereof? Were pain, as fuch, eligible, and pleafure, as fuch, avoidable; rewards and punifhments could be no motives to a man, to make him do or forbear any action. But if pleafure and pain have a neceffary effect on men, and if it be impoffible for men not to chufe what feems good to them, and not to avoid what feems evil; the neceffity of rewards and punifhments is then evident, and rewards will be of ufe to all thofe who conceive thofe rewards to be pleafure, and punifhments will be of ufe to all thofe who conceive them to be pain: and rewards and punifhments will

Human Liberty.　89

will frame thofe mens wills to obferve, and not tranfgrefs the laws.

Befides, fince there are fo many robbers, murderers, whoremafters, and other criminals, who notwithftanding the punifhments threatn'd, and rewards promis'd, by laws; prefer breaking the laws as the greater good or leffer evil, and reject conformity to them as the greater evil or leffer good: how many more would there be, and with what diforders would not all focieties be fill'd, if rewards and punifhments, confider'd as pleafure and pain, did not determine fome mens wills, but that, inftead thereof, all men could prefer or will punifhment confider'd as pain, and reject rewards confider'd as pleafure? men would then be under no reftraints.

VI. My fixth and laft argument to prove man a neceffary agent is: *Sixth argument is taken if*

90   *An Inquiry concerning*

from the if man was not a necessary agent
nature of determin'd by pleasure and pain, he
morality. would have no notion of *morality*, or
motive to practise it: the distinction
between morality and immorality,
virtue and vice would be lost; and
man would not be a moral agent.

Locke's   Morality or Virtue, consists in such
Essay of actions as are in their own nature,
H. Und. and upon the whole, *pleasant*; and im-
III. c. 20. morality or vice, consists in such acti-
Serjeant's ons as are in their own nature, and
Solid Phi-
lof. affert- upon the whole *painful*. Wherefore a 70
ed, *p. 215.* man must be affected with pleasure
and pain, in order to know what
morality is, and to distinguish it from
immorality. He must also be affected with pleasure and pain, to have a
reason to practise morality; for there
can be no motives, but pleasure and
pain, to make a Man do or forbear
any action. And a man must be
the more moral, the more he understands or is duly sensible, what actions

Human Liberty.   91

ons give pleasure and what pain; and
must be perfectly moral, if necessarily determin'd by pleasure and pain,
rightly understood and apprehended.
But if man be *indifferent* to pleasure
and pain, or is not duly affected
with them; he cannot know what
morality is, nor distinguish it from
immorality, nor have any motive
to practise morality, and abstain
from immorality; and will be equally indifferent to morality and immorality, or virtue and vice. Man
in his present condition is sufficiently
immoral by mistaking pain for pleasure, and thereby judging, willing,
and practising amiss: but if he was
indifferent to pleasure and pain, he
would have no rule to go by, and
might never judge, will, and practise
right.

Tho' I conceive I have so propo-   Objecti-
sed my arguments, as to have obvi- ons an-
ated most of the plausible objections swer'd.
usually

92   *An Inquiry concerning*

usually urg'd against the doctrine of necessity; yet it may not be improper to give a particular solution to the principal of them.

First ob-
jection.
* Auli
Gellii no-
ctes Att.
l. 6. c.:

1. First then it is objected, that * *if men are necessary agents and do commit necessarily all breaches of the law, it would be unjust to punish them for doing what they cannot avoid doing.* 71

Answer. To which I answer, that the sole end of punishment in society is to prevent, as far as may be, the commission of *certain* crimes: and that 72 punishments have their designed effect two ways; first, by restraining or cutting off from society the *vicious* members; and secondly, by correcting men or terrifying them from the commission of those crimes. Now let punishments be inflicted with either of these views, it will be manifest, that no regard is had to any *free-agency* in man, in order to render those punishments just; but

Human Liberty.   93

but that on the contrary punishments may be *justly* inflicted on man tho' a necessary agent. For, first, if *murderers* for example, or any such *vicious* members are cut off from society, merely as they are publick nusances, and unfit to live among men; it is plain, they are in that case so far from being consider'd as *free-agents*, that they are cut off from society as a canker'd branch is from a tree, or as a mad dog is kill'd in the streets. And the punishment of such men is *just*, as it takes mischievous members out of society. Also for the 73 same reason, *furious madmen*, whom all allow to be necessary agents, are in many places of the world, either the objects of judicial punishments, or are allow'd to be dispatch'd by private men. Nay, even *men infected with the plague*, who are not voluntary agents and are guilty of no crime, are sometimes thought to be justly cut

juftice? Whereas, a criminal who is an involuntary agent, (as for inftance, a man who has kill'd another in a chance medly or while in a fever, or the like) cannot ferve for an example to *deter* any others from doing the fame; he being no more an intelligent agent in doing the crime, than a houfe is, which kills a man by its fall: and by confequence the punifhment of fuch an involuntary agent would be unjuft. When therefore a man does a crime *voluntarily*, and his punifhment will ferve to deter others from doing the fame, he *is juftly punifh'd for doing what* (thro' ftrength of temptation, ill habits, or other caufes) *he could not avoid doing.*

It may not be improper to add this farther confideration from the law of our country. There is one cafe, wherein our law is fo far from requiring, that the perfons punifh'd fhould

cut off from fociety, to prevent contagion from them. Secondly, let punifhments be inflicted on fome criminals with a view to terrify, it will appear that in inflicting punifhments with that view, no regard is had to any *free-agency* in man, in order to make thofe punifhments *juft.* To render the punifhment of fuch men *juft*, it is fufficient that they were *voluntary* agents, or had the will to do the crime for which they fuffer for the law very juftly and rightly regardeth only the will, and no other preceding caufes of action. For example, fuppofe the 74 law on pain of death forbids theft, and there be a man who by the ftrength of temptation is neceffitated to fteal, and is thereupon put to death for it; doth not his punifhment *deter* others from theft? Is it not a caufe, that others fteal not? Doth it not frame their wills to juftice?

whose wills must be determin'd by it. It is as useful to such men, as the sun is to the ripening the fruits of the earth, or as any other causes are to produce their proper effects; and a man may as well say the sun is useless, if the ripening the fruits of the earth be necessary, as say, there is no need of threatning punishment for the use of those to whom threatning punishment is a necessary cause of forbearing to do a crime. It is also of use to society to *inflict* punishments on men *for doing what they cannot avoid doing*, to the end that necessary causes may exist, to form the wills of those who in virtue of them necessarily observe the laws; and also of use to cut them off as *noxious members* of society.

Second answer.

2. But secondly, so far is threatning and inflicting punishments from being useless, if men are necessary agents, that it would be useless

H

to

should be *free-agents*, that it does not consider them as voluntary agents, or even as guilty of the crime for which they suffer: so little is *free-agency* requisite to make punishments just. The children of rebel-parents suffer in their fortunes for the guilt of their parents; and their punishment is deem'd just, because it is suppos'd to be a means to prevent rebellion in parents.

75

Second objection.

II. Secondly, it is objected, *that it is useless to threaten punishment or inflict it on men to prevent crimes, when they are necessarily determin'd in all their actions.*

76

First answer.

1. To which I answer, first, that *threatning of punishments is a cause* which necessarily determines some mens wills to a conformity to law and against committing the crimes to which punishments are annex'd; and therefore is useful to all those whose

An Inquiry concerning

98

to *correct* and *deter* (which are the principal effects design'd to be obtained by threatning and inflicting punishments) unless men were necessary agents, and were determin'd by pleasure and pain; because, if men were free or indifferent to pleasure and pain, pain could be no motive to cause men to observe the law.

77

Third answer.

3. Thirdly, men have every day examples before them of the usefulness of punishments upon some intelligent or sensible beings, which they all contend are necessary agents. They punish dogs, horses, and other animals every day with great success, and make them leave off their vicious habits, and form them thereby according to their wills. These are plain facts, and matters of constant experience, and even confirm'd by the evasions of the advocates of liberty, who call *the rewards and punish-*

Bramball's Works, p. 65.

ments

Human Liberty.

99

78 ments us'd to brute beasts *analogical*; and say, that *beating them and giving them victuals, have* only *the shadow of rewards and punishments.* Nor are capital punishments without their use among beasts and birds. RORARIUS tells us, that *they crucify lyons in Africa* brute a-nim.&c. to *drive away other lyons from their cities* L 2. p.109. *and towns*; and that *travelling thro' the country of* Juliers, he *observ'd, they hang-* Quod

79 *ed up wolves to secure their flocks.* And in like manner with us, men hang up *crows* and *rooks* to keep birds from their corn, as they hang up murderers in chains to deter other murderers. But I need not go to brutes for examples of the usefulness of punishments on necessary agents. Punishments are not without effect on *some idiots* and *madmen*, by restraining them to a certain degree; and they are the very means by which the minds of *children*, are form'd by their parents. Nay, punishments have

H 2

100 *An Inquiry concerning*

have plainly a better effect on *chil-dren*, than on grown persons; and more easily form them to virtue and discipline, than they change the vicious habits of grown persons, or plant new habits in them. Wherefore the Objectors ought to think punishments may be threatned and inflicted on men usefully, thô they are necessary agents.

Third Objection.
3. Thirdly, it is objected, *if men are necessary agents, it is of no use to represent reasons to them, or to entreat them, or to admonish them, or to blame them, or to praise them.*

Answer.
To which I answer, that all these, according to me, are necessary causes to determine certain mens wills to do what we desire of them; and are therefore useful, as acting on such necessary agents to whom they are necessary causes of action; but would be of no use, if men had *free-will,*

Human Liberty. 101

will, or their wills were not mov'd by them. So that they who make this objection must run into the absurdities of saying, that *that cause is useful, which is no cause of action, and serves not to change the will; and that that cause is useless, which necessitates the effect.*

Let me add something further in respect of praise. Men have at all times been prais'd for actions judg-ed by all the world to be neces-sary. It has been a standing me-thod of commendation among the Epick Poets, who are the greatest Panegyrists of glorious actions, to attribute their Hero's valour and his great actions, to some Deity present with him and assisting him. Ho-MER gives many of his Hero's a God or a Goddess to attend them in bat-tle, or to be ready to help them in distress. VIRGIL describes ÆNEAS as always under the divine direction

H 3
and

102   *An Inquiry concerning*

and assistance. And TASSO gives the Christians in their holy war divine assistance.

Orators also and Historians, think necessary actions the proper subjects of praise. CICERO, when he maintain'd, that the Gods inspir'd MILO with the design and courage to kill CLODIUS, did not intend to lessen the satisfaction or glory of MILO, but on the contrary to augment it. But can there be a finer commendation than that given by VELLEIUS PATERCULUS to CATO, that he *was good by nature, because he could not be otherwise?* For, that alone is true goodness which flows from disposition, whether that disposition be natural or acquir'd. Such goodness may be depended on; and will seldom or ever fail. Whereas goodness founded on any reasonings whatsoever, is a very precarious thing; as may be seen by the lives of the greatest Declaimers against vice,

Oratio pro Milo- ne.

82

83

Human Liberty.   103

vice, who tho' they are constantly acquainting themselves with all the topicks that can be drawn from the excellency of goodness or virtue, and the mischiefs of vice; the rewards that attend the one, and the punishments that attend the other: yet are not better, than those, who are never conversant in such topicks. Lastly, the common proverb, *gaudeant bene nati*, is a general commendation of men for what plainly in no sense depends on them.

4. Fourthly, it is objected, that *if all events are necessary, then there is a period fix'd to every man's life: and if there is a period fix'd to every man's life, then it cannot be shortned by want of care or violence offer'd, or diseases; nor can it be prolong'd by care or physick: and if it cannot be shortned or prolong'd by them, then it is useless to avoid or use* [84] *any of these things.*

Fourth Objection.

H 4   In

Human Liberty. 105

whatsoever. For example, let it be fix'd and necessary for the river *Nile* annually to overflow; the means to cause it to overflow, must no less necessarily precede. And as it would be absurd to argue, that *if the overflowing of the Nile was annually fix'd and necessary, it would overflow, tho' the necessary means to make it overflow did not precede*; so it is no less absurd to argue from the fix'd period of human life, against the necessary means to bring it to its fix'd period, or to cause it not to exceed that period.

5. Fifthly, it is ask'd, *how a man can act against his conscience, and how a man's conscience can accuse him, if he knows he acts necessarily, and also does what he thinks best when he commits any sin?* I reply, that conscience being a man's own opinion of his actions with relation to some rule, he may at the time of doing an action contrary

Fifth Objection.

Answer.

104 An Inquiry concerning

Answer.

In answer to which, I grant, that if the period of human life be fix'd (as I contend it is) it cannot but happen at the time fix'd, and nothing can fall out to prolong or shorten that period. Neither such want of care, nor such violence offer'd, nor such diseases can happen as can cause the period of human life to fall short of that time; nor such care, nor physick be us'd, as to prolong it beyond that time. But tho' these cannot so fall out, as to shorten or prolong the period of human life; yet being necessary causes in the chain of causes to bring human life to the period fix'd, or to cause it not to exceed that time, they must as necessarily precede that effect, as other causes do their proper effects; and consequently when us'd or neglected, serve all the ends and purposes, that can be hop'd for or fear'd from the use of any means, or the neglect of any means what-

trary to that rule, know that he breaks that rule; and consequently act with reluctance, thô not sufficient to hinder the action. But after the action is over, he may not only judge his action to be contrary to that rule; but by the absence of the pleasure of the sin, and by finding himself obnoxious to shame, or by believing himself liable to punishment, he may *really accuse himself*; that is, he may condemn himself for having done it, be sorry he has done it, and wish it undone, because of the consequences that attend it.

Sixth Objection. 6. Sixthly, it is objected, *that if all events are necessary, it was as impossible (for example) for* JULIUS CÆSAR *not to have died in the Senate, as it is impossible for two and two to make six. But who will say that the former was as impossible as the latter is, when we can conceive it possible for* JULIUS CÆSAR *to have died*

any

any where else as well as in the Senate, and impossible to conceive two and two ever to make six?

To which I answer, that I do allow, *that if all events are necessary, it was as impossible for* JULIUS CÆSAR *not to have died in the Senate, as it is impossible for two and two to make six*: and will add, that it is no more possible to conceive the death of JULIUS CÆSAR to have happen'd any where else but in the Senate, than that two and two should make six. For whoever does conceive his death possible any where else, supposes other circumstances preceding his death than did precede his death. Whereas let them suppose all the same circumstances to come to pass that did precede his death; and then it will be impossible to conceive (if they think justly) his death could have come to pass any where else, as they conceive it impossible for two and two

to

An Inquiry concerning

to make fix. I observe also, that to suppose other circumstances of any action possible, than those that do precede it, is to suppose a contradiction or impossibility: for, as all actions have their particular circumstances, so every circumstance preceding an action, is as impossible not to have come to pass, by virtue of the causes preceding that circumstance, as that two and two should make fix.

The Opinions of the learned concerning liberty, &c. Having, as I hope, prov'd the truth of what I have advanc'd, and answer'd the most material objections that can be urg'd against me; it will, perhaps, not be improper to give some account of the sentiments of the learned in relation to my subject, and confirm by *authority* what I have said, for the sake of those with whom *authority* has weight in matters of speculation.

The questions of *liberty*, *necessity*, and *chance*, have been subjects of dispute among Philosophers at all times; and most of those Philosophers have clearly asserted *necessity*, and deny'd *liberty* and *chance*.

The questions of *liberty* and *necessity*, have also been debated among Divines in the several ages of the Christian Church, under the terms of *free-will* and *predestination*: and the Divines who have deny'd *free-will* and asserted *predestination*, have inforc'd the arguments of the Philosophers, by the consideration of some doctrines peculiar to the Christian Religion. And as to *chance*, *hazard*, or *fortune*, I think, Divines unanimously agree, that those words have no meaning.

Some Christian communions have even proceeded so far in relation to these matters, as to condemn in Councils and Synods the doctrine

The

110   An Inquiry concerning

of *free-will* as heretical; and the denial thereof is become a part of the *Confessions of Faith*, and *Articles of Religion* of several Churches.

From this state of the fact, it is manifest, that whoever embraces the opinion I have maintain'd, cannot want the *authority* of as many learned and pious men, as in embracing the contrary.

But considering how little men are mov'd by the *authority* of those who professedly maintain opinions contrary to theirs, thô at the same time they themselves embrace no opinion but on the *authority* of somebody; I shall wave all the advantages that I might draw from the *authority* of such Philosophers and Divines as are undoubtedly on my side: and for that reason shall not enter into a more particular detail of them; but shall offer the *authority* of such men, who profess to maintain *liberty*.

There

Human Liberty.   111

There are indeed very few real adversaries to the opinion I defend among those who pretend to be so; and upon due inquiry it will be found, that most of those who assert *liberty* in words, deny the thing, when the question is rightly stated. For proof whereof, let any man examin the clearest and acutest authors who have written for *liberty*, or discourse with those who think *liberty* a matter of experience, and he will see, that they allow, that the *will*, *follows the judgment of the understanding*; and *that, when two objects are presented to a man's choice, one whereof appears better than the other, he cannot choose the worst*; that is, cannot choose *evil* as *evil*. And since they acknowledge these things to be true, they yield up the question of *liberty* to their adversaries, who only contend, that the will or choice is always determin'd by what seems best. I will give

112 An Inquiry concerning

give my reader one example thereof in the most acute and ingenious Dr. CLARKE, whose authority is equal to that of many others put together, and makes it needless to cite others after him. He asserts, [87] that *the will is determin'd by motives*, and calls the *necessity*, by which a man chooses in virtue of those motives, *moral necessity*. And he explains himself with his usual candor and perspicuity by the following instance. *A man, says he, intirely free from all pain of body and disorder of mind, judges it unreasonable for him to hurt or destroy himself; and, being under no temptation or external violence, he* CANNOT POSSIBLY ACT *contrary to this judgment; not because he wants a natural or physical power so to do, but because it is absurd and mischievous, and morally impossible for him to choose to do it. Which also is the very same reason, why the most perfect rational creatures,*

Demonst. of the Being and Attributes of God, p. 105, of the 4th Edition 1716.

113 Human Liberty.

superior to men, cannot do evil; not because they want a natural power to perform the material action, but because it is morally impossible, that with a perfect knowledge of what is best, and without any temptation to evil, their will should determine itself to choose to act foolishly and unreasonably.

In this he plainly allows the *necessity*, for which I have contended. For he assigns the same *causes* of human *actions* that I have done; and extends the *necessity* of human actions as far, when he asserts, that a man *cannot* under those causes, *possibly do the contrary to what he does*; and particularly, that *a man* under the circumstances, of judging it *unreasonable* to hurt or destroy himself, and being under no temptation or external violence, *cannot possibly act contrary to that judgment*. And as to *a natural or physical power* in man to act contrary to that judgment,

and
I

and to *hurt* or *destroy* himself, which is asserted in the foregoing passage; that is so far from being inconsistent with the doctrine of *ne-cessity*, that the said *natural power to do the contrary*, or *to hurt* or *destroy* himself, is a consequence of the doctrine of *necessity*. For, if man is *necessarily* determin'd by particular *moral causes*, and *cannot then possibly act contrary* to what he does; he must under opposite *moral causes*, have *a power to do the contrary*. Man, as determin'd by *moral* causes, *cannot possibly* choose evil as evil and by consequence chooses *life* before *death*, while he apprehends *life* to be a *good*, and *death* to be an *evil*; as, on the contrary, he chooses *death* before *life*, while he apprehends *death* to be *a good*, and *life* to be *an evil*. And thus *moral causes*, by being different from one another or differently understood,

do

do determine men differently; and by consequence suppose *a natural power* to choose and act as different-ly, as those causes differently deter-mine them.

If therefore men will be govern'd by *authority* in the questions before us, let them sum up the real asser-ters of the *liberty* of man, and they will find them not to be very nume-rous; but on the contrary, they will find far the greater part of the pre-tended asserters of *liberty*, to be real asserters of *necessity*.

I shall conclude this Discourse with observing; that tho' I have contended, that *Liberty from Necessity* is contrary to experience; that it is impossible; and if possible, that it is an imperfection; that it is in-consistent with the divine perfecti-ons; and that it is subversive of laws and

The Author's notion of *Liberty from Necessity*.

I 2

Human Liberty.

He has also the same power in relation to the actions of his mind, as to those of his body. If he wills or pleases, he can think of this or that subject; stop short or pursue his thoughts; deliberate or defer deliberation or resume deliberation as he pleases; resolve or suspend his resolution as he pleases; and, in fine, can every moment change his object when he pleases: unless prevented by pain, or a fit of an apoplexy; or some such intervening restraint and compulsion.

And is it not a great perfection on in man to be able, in relation both to his thoughts and actions, to do as he wills or pleases, in all those cases of pleasure and interest? Nay, can a greater and more beneficial power in man be conceiv'd, than

to

An Inquiry concerning

and morality: yet, to prevent all objections to me, founded on the equivocal use of the word *Liberty*, which like all words employ'd in debates of consequence has various meanings affix'd to it, I think myself oblig'd to declare my opinion, that I take *man* to have a truly valuable *liberty* of another kind. He has a *power to do as he wills, or pleases.* Thus, if he wills, or pleases to speak or be silent; to sit or stand; to ride or walk; to go this way or that way; to move fast or slow; or, in fine, if his will changes like a weather-cock, he is able to do as he wills or pleases: unless prevented by some restraint or compulsion, as by being gagg'd; being under an acute pain; being forc'd out of his place; being confin'd; having convulsive motions; having lost the use of his limbs; or such-like causes.

He

118   *An Inquiry concerning,* &c.

to be able to do as he wills, or pleases? And can any other *liberty,* be conceiv'd beneficial to him? Had he this power or *liberty* in all things, he would be omnipotent!

F I N I S.

1. There is no indication in Collins' life or letters that this name was intended to refer to any particular person.
2. Cf. Locke, *Essay*, IV, 3, n. 1 and Introduction, n. 8, Fraser's edtition, II, p. 190, I, p. 32, for what Locke meant by the term "idea." Cf. also R. I. Aaron, *John Locke*, (Oxford, 1955), pp. 99–107 and F. Copleston, *A History of Philosophy*, (London, 1968), V, pp. 70, 71, for the ambiguity in his use of the term.
3. Cf. Locke, *Essay*, III, 2, n. 1 sq. Collins' note was added in the second edition of the *Inquiry*.
4. Cf. Locke, *Essay*, II, 31, n. 1 sq. for what Locke meant by "adequate" and "inadequate" ideas and II, 23, n. 33 sq. for his description of our idea of God.
5. Bayle, *Oeuvres*, IV, p. 862. Malebranche held that the ideas of things cannot come from the things themselves, but that God, who is intimately present to our souls, reveals to us the Divine Essence insofar as it is imitable in created beings. For a brief critique of the theory cf. Copleston, *A History*, (London, 1969), IV, p. 193 sq.
6. Cicero, *De Officiis*, I, 20, 70. Cf. Introduction p. 27.
7. J. de la Placette, *Eclaircissements sur quelques Difficultés qui naissent de la Considération de la Liberté*, (Amsterdam, 1709), p. 2. Jean de la Placette (1639–1718), who left France for Holland before the repeal of the Edict of Nantes, was placed in charge of the French church at Copenhagen by the Danish Queen Charlotte Amelia in 1686. He returned to Holland in 1712 and died there in 1718. He was regarded as one of the most outstanding protestant theologians of his time.
8. I. Jaquelot, *Dissertations sur l'Existence de Dieu*, (The Hague, 1697), p. 381. On p. 385 Jaquelot made it clear that he considered self-determination to be the essence of freewill. Bayle referred to Jaquelot and to this particular definition of freewill, *Oeuvres*, III, p. 798, *Réponse*, c. 145. Isaac Jaquelot (1647–1708) was in exile from France after the repeal of the Edict of Nantes. He was a preacher at The Hague, 1686–1691. He left The Hague after a quarrel with the extreme Calvinist Pierre Jurieu. In 1702 he became chaplain

to Frederick I of Prussia and declared himself openly as an Arminian. His *Dissertations* gave occasion to a protracted controversy with Bayle on freewill, the problem of evil and Pyrrhonism.

9. Alexander of Aphrodisias, *De Fato*, (London, 1658), p. 57. The *De Fato* was a defence of freewill against the determinism of the Stoics. Alexander lectured at Athens, A.D. 198–211. He was the most famous commentator on Aristotle but did not adhere rigidly to his doctrines. Cf. Copleston, *History*, (London, 1966), I, pp. 426, 427. The references to Alexander are in J. A. Fabricius, *Bibliotheca Graeca*, (Hamburg, 1711), IV, c. XXV, p. 62 sq. and J. G. Vossius, *De Philosophia et Philosophorum Sectis*, (The Hague, 1658), Vol. I, c. XVIII, sect. 8, p. 147.

10. The controversy can perhaps best be followed in the Molesworth edition of Hobbes' English Works, (London, 1839–). For this definition, cf. Vol. V, p. 361.

11. Hobbes, *English Works*, V, p. 253.

12. *Ibid.*, p. 279

13. *Ibid.*, p. 73 sq.

14. *Bibliothèque Choisie*, (Amsterdam, 1703–1718), Vol. XII, (1707), p. 102 sq. Jean Leclerc (1657–1736) was professor of Philosophy and Church History at the Remonstrant College at Amsterdam. His greatest influence was exercised through the literary journals of which he was editor, the *Bibliothèque Universelle et Historique* (Amsterdam, 1686–1693), the *Bibliothèque Choisie* (1703–1718) and the *Bibliothèque Ancienne et Moderne*, (1714–1730). For his connection with England, cf. R. Colie, *Light and Enlightenment*, (Cambridge, 1657), passim, for that with Collins, cf. J. O'Higgins, *Anthony Collins*, pp. 41, 77, 207, 214, 215. For his life, cf. Annie Barnes, *Jean Leclerc et la République des Lettres* (Paris, 1938).

15. Desiderius Erasmus, *Opera Omnia* (Leyden, 1706), tom. IX, col. 1215; *De Libero Arbitrio Diatribe sive Collatio.*

16. *Bibliothèque Choisie*, Vol. XII, p. 51.

17. Gilbert Burnet (1634–1715), Bishop of Salisbury after the Revolution of 1688. The citations are taken from *An Exposition of the Thirty-Nine Articles of the Church of England*, (London, 1699), pp. 117 and 27. In the latter passage he writes: "Thus there are such great difficulties on all hands in this matter, that it is much the wisest and safest course to adore what is above our apprehensions, rather than to enquire too curiously or determine too boldly in it." Page 27 is in the section dealing with Article I, on God and the Holy Trinity; page 117 in that on Article X, on man's freewill and the state of man after Adam's fall.

18. Bernadino Ochino (1487–1564), was successively General of the Observantine Franciscans and Vicar General of the Capuchins. However a series of sermons preached at Venice in 1539, which showed Lutheran tendencies, led ultimately to his flight from Italy. He became a Lutheran. From 1547 till 1553 he was in England, where he was made a prebendary of Canterbury. The *Labyrinth*, dedicated to Queen Elizabeth, was an attack on the

Calvinist doctrine of predestination. Collins' reference to Ochino is taken entirely from Bayle's *Dictionary*, article "Ochinus," *Dictionary*, IV, pp. 390–391, note P.

19. King, *De Origine Mali*, (London, 1702). The page references in Law's translation are respectively 228–229, 324, 245–249, 266–267 and 284–286. Collins' references are not always the best chosen for making his point.

20. M. S. Episcopius, *Opera Theologica*, (Amsterdam, 1650), part II, pp. 198–200. For the accuracy of Collins' interpretation of Episcopius cf. Introduction, p. 29. Simon Episcopius (1583–1643), was the systematiser of Arminianism. He studied under Arminius at Leyden, where he became professor of theology in 1612. He was the leader of the Arminian (or Remonstrant) representatives at the synod of Dort (1618–1619). After the victory of the Calvinist Counter-Remonstrants at the synod, he was exiled. He returned to Holland in 1626, when he became preacher at the Remonstrant College in Amsterdam.

21. J. Locke, *Some Familiar Letters between Mr. Locke and several of his Friends*, (London, 1708), p. 521, in a letter to the Arminian Philip van Limborch, of November 19th 1701. Collins' comment is rather more extreme than Locke's letters would seem to justify. Locke was asking Limborch for clarification on Episcopius' work and on Limborch's remarks on freewill in his last letter to him. Both had plainly left him with unanswered problems.

22. Again cf. Introduction p. 29 for the accuracy of the quotation. Anaxagoras (500ᶜ–428 B.C.), born at Clazomenae, had Pericles as one his of pupils. He is notable for being the first Greek philosopher to introduce the idea of mind as being the original cause of motion in the world, though he does not seem to have had a clear idea of the distinction between spiritual and material, and did not hold that mind is responsible for the order in the world.

23. William Reeves, *The Apologies of Justin Martyr, Tertullian and Minutius Felix*, (London, 1709), Vol. I, p. 159 (not p. 150. The 1709 edition was that in Collins' library – cf. Manuscript catalogue of the library of Anthony Collins, King's College, Cambridge, Keynes MS. 217, p. 379. This was the only edition printed in time to be used in the *Inquiry*). Reeves (1667–1726), fellow of King's College, Cambridge, chaplain to Queen Anne and Rector of Craneford in Middlesex, was attacking a sermon of the Calvinist divine, John Edwards (1637–1716), in which Edwards had revived the doctrine of predestination "in the rigid sense," which, said Reeves, "is not one jot better than fate in the sense of the Stoics." William Sherlock, in *A Discourse concerning the Divine Providence*, (London, 1694), p. 66, was also criticising predestination, which, he said, was a worse fate than any dreamed of by the ancient philosophers. Sherlock (1641ᶜ–1707), Dean of St. Paul's and Master of the Temple, 1685–1704, was a considerable controversialist. He aroused a good deal of criticism because, after defending passive obedience and opposing the succession of William and Mary, he took the oath of

allegiance in 1690. His High Anglicanism, therefore, did not lead him to become a non-Juror.

24. Bayle, *Dictionary*, III, p. 374, article "Helen," note Y. *Oeuvres*, IV, p. 726.
25. This is the gist of the whole chapter.
26. Leibniz, *Theodicy*, pp. 432, 434, in *Observations on the Book Concerning the Origin of Evil*.
27. *Journal des Savants*, (Amsterdam, 1705), tom. 33, p. 279.
28. In the edition in Collins' library that of 1706, (London), this passage occurs on pp. 3 and 4.
29. Cicero, *Academicorum Priorum*, II, 12, 38. The impression given by Collins is a little misleading. The passage in the dialogue was put in the mouth of Lucullus (110–57 B.C.), the conqueror of Mithridates. Actually it expressed the opinions, not of the Sceptics, but of Cicero's old master, Antiochus of Ascalon (135–68 B.C.), whose philosophy was eclectic and who was responsible for the end of Scepticism in the Academy. In the dialogues Lucullus was attacking the Sceptic position and, for the sake of the dialogue, Cicero (sect. 64 sq.) was defending it. The passage, therefore, can hardly be taken as an expression of the Sceptic point of view.
30. Sextus Empiricus, *Pyrrhoniarum Hypotyposes*, bk. I, c. 10, n. 19. There is a misprint in the text. Scepticism and Pyrrhonism (the holding of one's judgement in a state of suspense) became of considerable importance in the 16th and 17th centuries. The frequent combination of fideism and Pyrrhonism is exemplified in Pierre Bayle. The early history of the movement is discussed in Professor Popkin's, *The History of Scepticism from Erasmus to Descartes*, (Harper Torchbooks, New York, 1968). Pyrrhonism received its name from Pyrrho of Elis (360ᶜ–275ᶜ B.C.). The philosophers know as the Middle Academics – who came after Plato – and the New Academics were Sceptics. Cicero (106–43 B.C.) was an eclectic, inclined to Scepticism in natural philosophy but basing moral judgements on innate notions. Sextus Empiricus, in the late second century A.D., in his *Hypotyposes* and *Adversus Mathematicos* gives the fullest account of the doctrines of the Sceptics. Collins gives little sign of Scepticism, but in his *Essay on the Use of Reason* he wrote (p. 13): "I deny not the distinctions of real and seeming appearances ... yet notwithstanding that we can only govern ourselves by seeming relations and appearances, because real relations and appearances can but seem to be relations and appearances."
31. For this argument, cf. Introduction p. 30 sq.
32. Locke in this passage is not concerned with the question whether we can will what our judgement tells us is less attractive. What he is saying is that the will is not free – in the libertarian sense – because in the *actual moment* of choice one cannot be *at the same moment* willing something else. His argument is based on the idea that freedom in the will, in the libertarian sense, would demand that every volition be preceded by a previous volition and would therefore lead to a *progressus in infinitum*. A. C. Fraser's comment on the argument is that the assumption is unwarranted and an act

of will itself a first cause: *Essay*, Vol. I, p. 328, n. 3. In these passages of his *Essay* Locke defines freewill as freedom from external obstruction; cf. II, 21, n. 27.

33. John Norris, *The Theory and Regulation of Love*, (Oxford, 1688), in *Letters philosophical and moral between the Author and Dr. Henry More*, appended to *The Theory* . . ., p. 199. Norris (1659–1711), fellow of All Souls College, Oxford, was the youngest and the only Oxford member of the group known as the Cambridge Platonists. Henry More (1614–1687), fellow of Christ's College, Cambridge, was one of the leading members of the group. The Cambridge Platonists were very concerned with the question of freewill and took Thomas Hobbes as their particular opponent. In the passage referred to by Collins, Norris, in a reply to More, in his 4th letter, was defending his theory that the seat of freedom does not lie immediately in the soul as "volent" but "in the soul as intelligent" and in the power "to attend or not attend or to attend more or less" to the possible objects of choice. The "true and ultimate ground of all sin," he said, lies in the fact that "moral corruptions may divert the soul from sufficiently attending to the beauty of holiness," (pp. 201–204). Cf. also Introduction, p. 31.

34. Bayle, *Oeuvres*, III, p. 784; *Réponse*, c. 139.

35. The passage is taken from Plato, *Protagoras* 345, D and 358, C and D. In the *Protagoras* Plato can be interpreted as arguing for a form of hedonistic calculus, the identification of the pleasant and the good. However, it is not certain that this is his own view and it is not consistent with the view expressed in almost every other dialogue in e.g. the *Gorgias*, 497, A and D, in which he argues that there is a distinction between good and evil pleasures; cf. J. and A. Adams, *Protagoras*, (Cambridge, 1928), Introduction, p. XXX. Plato's own opinion on freewill is not completely clear. T. Gomperz, *Greek Thinkers*, (London, 1905), III, p. 258, says that "there are isolated phrases which seem to stamp Plato as an indeterminist." However, he says "this impression . . . will not bear scrutiny." He holds that Plato held to the intellectual determinisn of Socrates. On the other hand E. Zeller, *Plato and the Older Academy*, (London, 1888), p. 421, says that we are not justified in disregarding the enunciations on freewill that we find in Plato. He says, (p. 420), that Plato maintains with Socrates that no-one is voluntarily bad, but he adds that, in Plato's opinion, ignorance of what is truly good is still the man's own fault. He questions Plato's consistency in declaring all ignorance and wickedness involuntary and at the same time saying that man's will is free and man morally responsible. It may be best, with Zeller, to say that Plato was probably unconscious of the dilemma in which he was involved. In more recent works contrast J. M. Robinson, *Plato's Psychology*, (University of Toronto Press, 1970), p. 107 sq. and R. Demos, *The Philosophy of Plato*, (London, 1939), p. 333 sq.

36. Hobbes, *English Works*, V. pp. 67, 73. Cf. Introduction, p. 32.

37. For a comparison with Leibniz, cf. Introduction, pp. 23 sq., 32.

38. Aristotle, *Nicomachean Ethics*, VII, 3, 1147^a. Collins was using an older and

alternative notation that gives the reference as VII, 5. In the passage quoted, Aristotle was dealing with the actions of the incontinent man and asking how it is that he can act contrary to what "right reason," i.e. ethically right reason, demands. In general he rejected the Socratic position that sin is ignorance and he did hold that moral action demands freedom e.g. *Eudemian Ethics*, II, 7, 1223^a. At the same time he did not completely escape from the influence of Socrates and in the *Nicomachean Ethics*, bk. VII, in dealing with continence and incontinence he inclined, on one interpretation, to the idea that the incontinent man, doing a wrong act, does not know, when he is acting, that the act is wrong. The question is a controversial one. For a recent discussion, cf. R. D. Milo, *Aristotle on Practical Knowledge and Weakness of Will*, (The Hague, Paris, 1966).

39. Cf. Bayle. *Dictionary*, IV, p. 908, article "Rorarius," note F; Hobbes, *English Works*, IV, p. 244.
40. Hobbes, *English Works*, V, pp. 40, 66.
41. For Hobbes' argument on causality, cf. *English Works*, IV, p. 276.
42. Cf. Locke, *Essay*, IV, 10, n. 3, in Locke's argument for the existence of God.
43. Bayle, *Dictionary*, II, p. 790 sq., article "Epicurus," note U.
44. Lucretius, *De Rerum Natura*, II, line 250 sq.; Eusebius, *Preparatio Evangelica*, (Cologne, 1688), lib. VI, c. VII, especially p. 261; Cicero, *De Natura Deorum*, I, 23. 65 sq. especially 69, 70. Titus Lucretius Carus (94^c–55 B.C.) was the best-known member of the Epicurean School. He expounded the physical theory of Epicurus in his *De Rerum Natura*, aiming at freeing men from fear of the Gods and of the punishment of the soul after death. For the Stoic rejection of freewill, cf. E. Zeller, *The Stoics, Epicureans and Sceptics*, (London, 1880), p. 173 sq. For the Epicurean position, cf. *Ibid.*, p. 445 sq. and 459 sq. More recent works are J. Rist, *Stoic Philosophy*, (Cambridge, 1969), and D. J. Furley, *Two Studies in the Greek Atomists*, (Princeton, 1967).
45. Collins' references are taken according to the numeration given in Sir Roger L'Estrange's translation of Josephus' works, (London, 1702). In Whiston's translation the references are *Antiquities*, XVIII, c. 1 and *De Bello Judaico*, II, c. 8. Cf. Introduction, p. 34.
46. J. Stearne, *De Obstinatione, or concerning Firmness and not sinking under Adversities*, (Dublin, 1672). Sects. 40, 41 of Dodwell's *Prolegomena* to the work to deal with Paul, say he had a Pharisaic tendency, even as a Christian and that the Pharisees derived many of their ideas from the Stoics, but, with regard to fate, Dodwell says that both Stoics and Pharisees saved freewill; "Sicut enim Stoici, ita etiam Pharisaei, fatum ita explicabant ut salvum tamen esset liberum arbitrium humanum. Unde etiam de iis non semel testatus est Josephus ex eorum sententia, mixtionem quamdam esse ex consilio divino et arbitrio humano, quaedam etiam ex fato proficisci, quaedam ex libero solo hominum arbitrio." *Op. cit.*, p. 148. John Stearne (1624–1669), a doctor, was the founder of the Irish College of Physicians. His works were mainly theological. He left his *De Obstinatione* to be

published posthumously by his pupil Henry Dodwell, (1641–1711), fellow of Trinity College, Dublin, Camden professor of History at Oxford from 1688 till 1691, when he was deprived for refusing to take the oath of allegiance.

47. Cf. Introduction, p. 35.
48. In Law's translation, p. 333.
49. G. Cheyne, *Philosophical Principles of Religion, Natural and Revealed*, (London, 1715), Part I, c. III, sect. 13. Cf. Introduction, p. 35. Cheyne (1671–1743), like Stearne, was a physician. He studied at Edinburgh, lived in London and finally, for his health's sake, at Bath.
50. The example is given by Bayle, *Oeuvres*, III, p. 662; *Réponse*, c. 81.
51. Hobbes, *English Works*, V, pp. 59, 60.
52. In Law's translation, pp. 279, 280.
53. *Ibid.*, pp. 268–272.
54. Cf. Bayle, *Oeuvres*, III, p. 661; *Réponse*, c. 80.
55. In Law's translation, pp. 375–379.
56. *Ibid.*, pp. 267–268, 275–276.
57. *Ibid.*, pp. 267–268, 279–280, 349–350, 355–362, 373–374.
58. *Ibid.*, p. 286. 177 in the note in the text is a misprint for 117.
59. Cf. Leibniz, *Theodicy etc.*, p. 431.
60. Note the enunciation of the principle of causality. Cf. Locke, *Essay*, IV, 10, n. 3.
61. Cf. Leibniz, *Theodicy etc.*, p. 431.
62. In Law's translation, p. 286.
63. Burnet is guilty of theological unorthodoxy, verbally at least, in saying that the transient acts of God are done in a succession of time. The acts, as Divine acts, cannot be subject to time or succession. It is their external effects that occur in a succession of time. If God is a simple perfect being, his act of creation, for example, is one with his Essence and therefore necessary. It is, however, not subject to necessity, as God freely creates. The word "necessary" is not being used in the same sense in the two cases. His act, as being one with his Essence, is necessary, in that its existence is necessary; it is not subject to necessity in that neither external causes nor the internal constitution of the Divine Essence compel him to create. There remains, however, a mystery, as Burnet says, – how an act, which is not necessitated, should necessarily, de facto, exist from all eternity.
64. Hobbes, *English Works*, V, p. 247; cf. also Bayle, *Oeuvres*, III, p. 662; *Réponse*, c. 81.
65. This passage is taken verbatim from Bayle, *Oeuvres*, II, p. 679; *Réponse*, c. 90.
66. Hobbes, *English Works*, IV, pp. 270, 271, V, p. 328 sq.
67. Cicero, *De Divinatione*, bk. II, 6, 17 and 7, 18. In book II of *De Divinatione* Cicero is ridiculing the idea of divination. The passage as quoted by Collins is not easy to understand, as Collins omits Quintus, Cicero's brother's definition of divination – "The foreknowledge and foretelling of things that happen by chance" – which Cicero is using to reply, in the dialogue,

to one of Quintus' arguments, based on Stoic principles, that, although omniscience only belongs to God, a man, gifted with the power of divination, can sometimes perceive the future in its causes. Chance is opposed to fate, Cicero says. One can either accept chance and surrender the Stoic doctrine of fate, or reject chance and scrap Quintus' definition of divination. Cicero's own ideas, being those of an eclectic, are not easy to perceive, but it would seem from his very late work, *De Fato*, XI, 25, that he accepted the idea of freewill – or, at any rate, rejected the idea of choice enforced necessarily by external causes or circumstances. He did not discuss the question of psychic determinism.

68. M. Luther, *On the Bondage of the Will*, translation by J. I. Packer and O. R. Johnston (London, 1957), pp. 216, 217. Luther's work was not originally divided into chapters. In it he was not concerned primarily with freewill as such but with the theological question of merit. This section, in which he argues from reason, can be taken to have a wider application, and it supports Collins' point of view. Earlier Luther wrote: "The will, be it God's, or man's does what it does, good or bad, under no compulsion, but just as it wants or pleases, as if it were free. Yet the will of God ... is changeless and sure ... and our will, principally because of its corruption, can do not good of itself" (p. 81). He went on (p. 82), "Our original proposition still stands and remains unshaken: all things happen by necessity." Man's will, therefore, has spontaneity. In this sense it is free from compulsion. But it is necessitated, by its corruption, to act of itself according to its corrupt nature. As the translators say (p. 48), Luther's denial of freewill had nothing to do with the psychology of action. In spite of this, the passage quoted by Collins could have a general application.

69. R. South, *Twelve Sermons upon several Subjects*, (London, 1698), Vol. III, p. 487, 488. Robert South (1634–1716), was a notable and salty preacher, public orator at Oxford, 1660–1667, and then Rector of Islip. He declined the see of Rochester in 1713. In the passage quoted by Collins, taken in its full context, he distinguished between physical and causal necessity, which, he said, certainly caused an event, and what he called "logical" necessity, by which an event is with certainty inferred but by which it is not efficiently caused. (This is not the present day sense of the term "logical necessity.") This latter type of necessity, he considered, does not take away freewill and it is to this type, he thought, that God's foreknowledge belongs. Descartes, *Principia Philosophiae*, I, 41, said that by our self-consciousness we are quite sure that we are free, that we cannot see how this can be reconciled with God's foreknowledge, but that it would be absurd to doubt something of which we are so sure, because of a difficulty arising from something else – i.e. God's foreknowledge – which is incomprehensible to us.

Tillotson, *Sermons*, (London, 1700–1706), 2nd. ed., Vol. VI, p. 157, said the problem is "contradictious and impossible to us," but that we have "sufficient assurance" of the thing and would need infinite understanding to unravel it. Stillingfleet, *A Discourse concerning the Doctrine of Christ's*

Satisfaction, (London, 1697), p. 355 sq., said that there is something above our comprehension in the connection between the certainty of Divine prescience and the liberty of human actions. John Tillotson (1630–1694), Archbishop of Canterbury, 1691–1694, and Edward Stillingfleet (1635–1699), Bishop of Worcester, 1688–1699, were two of the leading Latitudinarian Divines, raised to the Episcopacy after the Revolution of 1688.

For Locke's opinion, cf. Introduction, p. 11. The quotation is taken from J. Locke, *Some Familiar Letters between Mr. Locke and several of his Friends*, (London, 1708), p. 27.

70. For Locke's view, cf. Introduction, p. 12. J. Sergeant's *Solid Philosophy asserted against the Fancie of the Ideists*, (London, 1697), was a criticism of Locke, but in this passage (p. 215 sq.), he considered he agreed with him. However, as he said that true morality depends on our choosing wisely, his morality was not subjective. What is truly good, he held, should be and was intended by God to be truly pleasurable. Sergeant (1622–1707), was a noted and rather controversial Catholic priest and controversialist.

71. The *Noctes Atticae* of the Roman grammarian, Aulus Gellius (A.D. 130–180), were a compilation of notes on a great variety of subjects and are of great use in providing excerpts from the lost texts of earlier writers. In this passage he was giving an objection raised against Chrysippus the Stoic by the defenders of freewill. The passage is in book VII, 2, 5, not book VI.

72. Cf. Hobbes, *English Works*, IV, p. 253.

73. *Ibid.*

74. *Ibid.* The following passage in Collins is taken almost verbatim from Hobbes.

75. Till the Forfeiture Act of 1870 (33 and 34 Vict. c. 23) a person convicted of High Treason was held in law to have his blood corrupted. There could be no inheritance claimed through corrupted blood and the Crown became absolutely entitled to the convict's land. Sect. 10 of the Inheritance Act of 1833 (3 and 4 Will. IV, c. 106) provided that, after the death of an attainted person, his descendants might trace their descent through him as though he had not been attainted. It is, therefore, a moot point whether, in Collins' time, the children were regarded, in law, as being punished directly, even though they suffered because of their parents' treason.

76. Cf. Hobbes, *English Works*, IV, p. 255. Leibniz, *Theodicy*, p. 160.

77. Cf. Leibniz, *Theodicy*, p. 347.

78. Hobbes, *English Works*, V, p. 172.

79. In the common, Paris, edition of 1648, pp. 145–146. Hieronymus Rorarius (1485–1556), Nuncio of Pope Clement VII at the court of Hungary, published his work in 1548. He argued that beasts are more intelligent than men. The theory had some favour in his own day. It was taken up by Bayle, who considered it absurd, but who used the examples Rorarius gave of animal intelligence, to the embarrassment of the Cartesians, who considered beasts to be machines. Cf. Bayle, *Dictionary*, III, p. 900 sq., article "Rorarius." The quotation used by Collins is given in note F, p. 908. Cf. also Leibniz, *Theodicy*. p. 160.

80. Cf. Hobbes, *English Works*, IV, pp. 248, 252, 254–256.
81. Cf. Leibniz, *Theodicy*, pp. 317, 318, referring to Bayle, *Oeuvres*, III, p. 658 sq.; *Réponse*, c. 80.
82. Cf. Hobbes, *English Works*, V, p. 53. Cicero, *Pro Milone*, 30, 83 sq.
83. Velleius Paterculus, *Historia Romana*, Lib. II, c. 35. This was a common example, given by Hobbes, *English Works*, IV, p. 256 and Leibniz, *Theodicy*, p. 318. Velleius Paterculus (19^c B.C. – 31^c A.D.), the Roman historian, wrote of Cato, called "of Utica," (95–46 B.C.), a leading opponent of Julius Caesar, who committed suicide at Utica, after Caesar's victory at the battle of Thapsus, that he never acted with rectitude in order to appear righteous, but because he could not do otherwise. The statement seems to justify Bramhall's comment, (Hobbes, *English Works*, V, p. 172), that "the true meaning is that he was naturally of a good temper, not so prone to some kinds of vice as others were." It does not seem to justify the extreme interpretation put on it by Hobbes, Leibniz and Collins.
84. Cf. Hobbes, *English Works*, IV, p. 256. It is worth noting how Priestley compared Collins and Hobbes. "The great merit of this piece," – the *Inquiry* – he wrote, "consists in its conciseness, its clearness, and its being the first regular treatise on the subject. Mr. Hobbes, I am still of opinion, was the first who, in this, or any other country, rightly understood, and clearly stated, the argument; but he wrote nothing systematical, and consequently nothing that could be of much use to a student." He added that there were "few topics in the whole compass of the argument, which he" – Collins – "has not touched upon." J. Priestley's edition of the *Inquiry*, (Birmingham, 1790), Preface, pp. III, IV.
85. For this significant objection, cf. Introduction, p. 41 sq.
86. The Reformed Churches considered that man's will had been corrupted since the fall of Adam, but they were less concerned with the philosophical question of freewill than with the theological question concerning meritorious acts. It is hard to see what evidence Collins could have produced for their going so far as to declare the philosophical belief in freewill to be heretical. For an outline of the official protestant declarations that touched on freewill cf. W. A. Curtis, *A History of Creeds and Confessions of Faith*, (Edinburgh, 1911), pp. 146, 159, 174, 186, 210, 216, 245, 271 and 368. Luther's opinion on freewill has already been dealt with, (sup. note 68). For Calvin, cf. F. Wendel, *Calvin*, (Fontana Library, London, 1965), pp. 188–193. Wendel says Calvin follows Luther in his *De Servo Arbitrio*. He quotes Calvin as saying that, since the fall, man "has not been deprived of will, but of healthy will" (p. 189), but he also says that "he prefers to follow Luther in denying freewill altogether; and like Luther in *De Servo Arbitrio* he defends the distinction between necessity and restraint" (p. 190). This doctrine could well fit in with Collins' psychic determinism. But it must be repeated that Luther was dealing with a theological and not a psychological problem.
87. Cf. Introduction, p. 42 sq.

COLLATION OF THE TEXT OF THE *INQUIRY*

The edition of the *Philosophical Inquiry* here reproduced is the second edition, (London, 1717), which is described as "The second edition corrected." However, there are a number of textual variations. These are:

In the "Contents" in the first edition the words "The *Introduction* ... all subjects" are omitted.

Under "I First Argument," 1.8: "*Perceiving*" is, in the first edition, "perception."

The word "consider'd" is appended, in the first edition, to each of the first four of "Several objections consider'd," and the word "answer'd" to objections 5 and 6.

p. 1   The title "A Philosophical Inquiry etc." is omitted, in the first edition, before the dedication "To Lucius."

p. 2   1.10: All the satisfaction I can. – 1st ed. A *second Discourse* on this subject.

p. 3   The footnote "I do not mean unknown simple ideas ..." is omitted in the first edition. Collins' intention in adding the footnote may have been to distinguish between mysteries and truths of which we are not aware because we have never experienced or heard of them, but which we can fully understand when we experience or are told of them.

p. 11   1.14: hath been – 1st ed., was

p. 11   1.20: not unavoidably – ist ed., not ever unavoidably

p. 14   1.9: these – 1st ed., those

p. 18   1.19: Whereas when the mind ... *moment* of action – 1st ed., "Whereas this indifference is a necessary state of mind, to which the mind is no less determin'd during its deliberation, than it is when it acts, or not acts, after it comes out of that state by the means of deliberation; and not the less necessary, *because the mind is not under an* actual *determination to act, or not to act*; which actual *determination* must ever be subsequent to a state of *indifferency*." The version in the second edition gives Collins' opinion in a simpler and, therefore, clearer way, and brings out the fact that he considered the mind to act as a sort of balance, with ideas and

"motives" as the weights. For Bayle's comparing the act of choice to weighing in a balance, and for Leibniz's and Clarke's opinions on the meaning of the word "motive," cf. Introduction, pp. 31, 23.

p. 20   1.11: and that therefore *he pretends* – 1st ed., and therefore *pretends*

p. 25   1.15: write obscurely – 1st ed., talk obscurely

p. 25   1.16: at least, he will see that – 1st ed., at least, that

p. 26   1.8: speaks thus: *Fate*, says he, – 1st ed., tells us, *That Fate*

p. 28   1.8: *That the best proofs alledged for liberty are,* that without it, – 1st ed., That *liberty cannot be proved from experience:* And *that the best proofs thereof are,* that *without liberty,*

p. 28   1.11: *evil, as well as good thoughts.* – 1st ed., *evil thoughts, as well as good*

p. 29   1.20: The Journalists ... they say, – 1st ed., The Journalists of *Paris* pass this censure on the aforesaid notion of *Liberty.* Mr. King say they,

p. 30   1.15: the vulgar experience – 1st ed., that vulgar experience

p. 32   1.17: is the foundation – 1st ed., to be the foundation

p. 32   1.19: makes them – 1st ed., to make them

p. 35   1.2: *adherence to which,* – 1st ed., *adherence to,*

p. 42   1.4: capable of willing or preferring, – 1st ed., capable of willing or preferring, or chusing

p. 45   1.22: for the sake whereof ... absurd – 1st ed., for the sake whereof there is so much contest for so absurd

p. 52   1.3: preferable – 1st ed., preferably

p. 53   1.3: an action – 1st ed., the action

p. 60   1.9: the Jews, I say, who besides – 1st ed., who besides

p. 69   1.21: and *reason,* determine – 1st ed., *reason* do determine

p. 77   1.10: *his knowledge and decrees,* – 1st ed., *his knowledge and his decrees*

p. 87   1.4: *the thing* – 1st ed., *that thing*

p. 90   1.8: "Morality ... painful." In the first edition the words "and upon the whole" are omitted, with regard to both pleasant and painful actions. The inclusion, – perhaps a qualification – in the second edition, in no way alters the hedonism of Collins' theory of morality.

p. 97   1.21: So far is – 1st ed., So far are

p. 97   1.23: if men are – 1st ed., if men were

p. 97   1.24: that it would be useless – 1st ed., that they would be useless
The passage "to correct ... inflicting punishments)" is omitted in the first edition. Its inclusion in the second emphasises the fact that Collins regarded correction and deterrence as the principal purposes of punishment. Elsewhere he said that prevention of certain crimes is its sole purpose (p. 92 and Introduction, p. 38 sq.). "Correction," therefore, seems to be meant to carry the sense of deterring the criminal for the future, if Collins is to be taken as being consistent.

p. 107 1.20: be impossible to conceive – 1st ed., be as impossible to conceive

p. 108 1.17: it will, perhaps, not be improper – 1st ed., it may be proper

p. 109 1.5: clearly – 1st ed., "ly", corrected in the *errata* to "openly"

p. 115　1.20: "that it is inconsistent with the divine perfections" is omitted in the first edition.

p. 117　1.4: of this – 1st ed., on this

VARIATIONS IN THE MARGINAL NOTES GIVEN BY COLLINS

p. 17　1.3: 1st ed. adds 715.

p. 21　King de Orig. Mali p. 91, 127. – 1st ed., King de Orig. Mali 127, 91

p. 27　Dictionnaire etc – 1st edition omits "2d edit."

p. 29　Remarques etc. p. 76. – 1st ed. Remarques etc., 79

p. 30　Journal des Savans etc. – 1st edition simply has "Mois de Mars, 1705."

p. 31　1st edition omits "Perception of Ideas."

p. 33　1st edition omits "Judging of Propositions."

p. 36　1st edition omits "Willing."

p. 38　1st edition adds to "Locke of Hum. Und. 1.2. c21." "King de Orig. Mali. 101."

p. 42　Theory of Love etc. – 1st edition omits p. 199.

p. 43　The first edition, in the reference to Plato, omits "Edit. Serran." and has, instead of "345, 346," "345, 358."

p. 44　Bramhall's works, p. 656 and 658. – 1st ed., Bramhall's works 65, 658

p. 52　1st edition omits "Doing as we will"

p. 76　pag. 117 – 1st ed., 117

p. 112　of the 4th *Edition* 1716. – 1st ed., of the *Edition* 1716

The text of the *Philosophical Inquiry* is reproduced by the kind permission of the President and Fellows of Corpus Christi College, Oxford.

De licentia Superiorum Ordinis.

Printed in the United States
117374LV00001B/200/A